Math Competitions
in South Texas
AND SOME MAGIC TRICKS

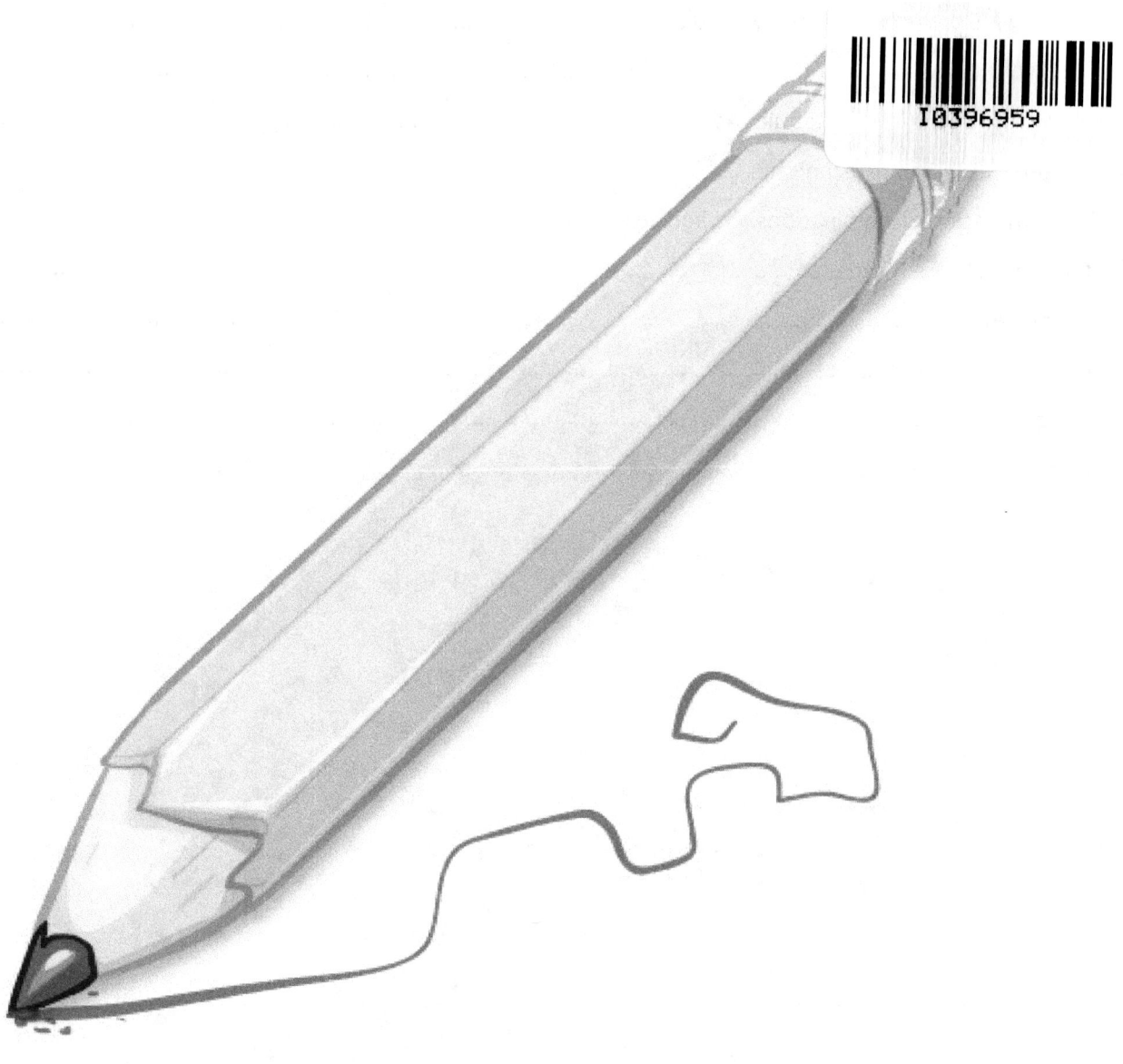

Dr. Ricardo Teixeira

Contents

Chapter 1: Algebra Problems 5
    Simple Algebra 5
Chapter 2: Graphs 22
    Reading graphs and getting basic graphs on the plane 22
Chapter 3: Geometry 29
    Measures of angles, of lengths and areas 29
Chapter 4: Number Theory 38
    Divisibility Rules and Modular Addition 38
Chapter 5: Counting Techniques 44
    Permutations, Combinations, Probability, Counting Techniques 44
Chapter 6: More Math 55
    Patterns 55
Chapter 7: "Mathemagics" 65
Appendices 74
About the author 82

Introduction

Math and Robotics Awareness Day

These notes were inspired by many years organizing a Math event at the University of Houston – Victoria (UHV). Annually, hundreds of high-school students attend a one-day event at UHV campus, and they participate in math competition, and other activities related to math and robotics. The name of the event is "Math and Robotics Awareness Day", and it coincides with the Math Awareness Month: April of each year.

Mathematics and Computer Science faculty members, together with UHV staff, prepare a day (from 8am to 2pm) of several activities. The organization of the event takes months, and it is a really detailed process. We have to recognize Paula Edging and Tracey Moore-Sweeney for their outstanding leadership on the event.

A typical day would have registration; an introduction by UHV dean, provost or president; math competition; computer science competition; robotics demonstrations; exposition booths from local industries and businesses; talks from UHV professors; a *mathemagic* demonstration by this author; games; lunch; digital gaming lab tour; and the award ceremony to close the day.

Most funds for the event come from a partnership with ALCOA.

With time, our event grew in size and in quality. Local teachers started to ask for materials to better prepare their students. However, we could not find a good textbook to recommend. The only thing we were able to do was to publish some practice questions, and previous tests. After few years, we realized we had accumulated enough material to create our own book.

In this note, you will find ALL problems from six editions of the competitions: 2011 to 2016. Also, we answered the practice questions that were published for preparation.

You will also note that some questions were repeated on different editions, or were on practice material and were later used on a competition. This was intentional, so students and teachers would attempt to do them.

Content

The <u>questions do not require deep knowledge of mathematics</u>. There is no Calculus or more abstract concepts. Most students in high school, even freshmen, would have been already exposed to material on most of the questions.

We divide the book into chapters:

- **Chapter 1:** it covers algebra problems. Main topics are computational manipulation, percentages, quadratic equations, exponent rules, simple equations, simple word problems, systems of equations, proportionally and inversely proportionally problems, and more.
- **Chapter 2:** questions involving graphs. We will discuss how to interpret graphical data, and some questions on lines and parabolas.
- **Chapter 3:** some geometry problems. The central theme of this chapter includes measures of angles, of lengths (Pythagorean Theorem) and areas.
- **Chapter 4:** number theory. We summarize divisibility rules for several numbers, and present really beautiful questions regarding it. We also cover Modular Addition and some clock calculations.
- **Chapter 5:** a chapter on counting techniques. Besides the questions, chapter includes a brief explanation about permutations, combinations, probability, and other counting techniques.
- **Chapter 6:** patterns and more. Several different types of questions in which the identification of a pattern may be crucial. Plus, there are some other problems that quite did not fit on previous chapters.
- **Chapter 7:** some magic tricks. A chapter in which I summarize some magic tricks that could be explained via different base number systems.

Chapter 1: Algebra Problems

Simple Algebra

Some of the questions only require simple algebra. Remember that students are not allowed to use calculators on the exams, but also no intense computation is intended.

Questions will test basic skills such as computational manipulation, percentages, quadratic equations, exponent rules, simple equations, simple word problems, systems of equations, proportionally and inversely proportionally problems, and more.

For the computational manipulation, we test the order of operations. First, the operations inside <u>P</u>arenthesis must be done. If there is any <u>E</u>xponents, then they have to be done afterwards. <u>M</u>ultiplication and <u>D</u>ivision have the next priority. And the last operations to be done are <u>A</u>ddition and <u>S</u>ubtraction. This is what teachers like to refer as PEMDAS. We also test operations with simple fractions.

Percentage: simply consider the symbol % to be the same as "multiplication by $\frac{1}{100}$". A good hint is that 10% is the same as $10 \times \frac{1}{100} = \frac{1}{10}$. Hence, 10% is simply one tenth.

Quadratic equations: although factoring is exhaustively taught in high schools, the best way to solve a complete quadratic equation $ax^2 + bx + c = 0$, with a, b, c not zero, is using the quadratic formula: $x = \frac{(-b \pm \sqrt{b^2 - 4ac})}{2a}$ the parentheses are to enforce that $2a$ is dividing the entire expression. If either b or c is zero, we could still use the above formula, but we could solve the equation quicker by:

- If $b = 0$: then the general equation becomes $ax^2 + c = 0$. Then $ax^2 = -c$. So, the answers would be $x = \pm\sqrt{-\frac{c}{a}}$.
- If $c = 0$: then $ax^2 + bx = 0$. Putting x as a common factor, $x(ax + b) = 0$. The product between two expressions is zero only if one of the terms is zero. So, either $x = 0$ or $ax + b = 0$, which makes $x = -\frac{b}{a}$.

The main exponent rules are (consider the "base" b to be a positive number, a and c could be any real number): multiplication with same base $b^a \times b^c = b^{a+c}$, division with same base $\frac{b^a}{b^c} = b^{a-c}$, power of power $(b^a)^c = b^{ac}$.

For the rest of problems, explanations are self-sufficient.

1. (Competition 2016) What is $7 - 4 + 5 \times 0 + 1$?

   Answer: 4

   Solution: We need to observe that multiplication has a higher "priority" than addition and subtraction: $7 - 4 + 5 \times 0 + 1 = 7 - 4 + 0 + 1 = 4$.

2. (Competition 2011) Evaluate $\dfrac{3}{4} + \dfrac{5}{6}$:

   A. $\dfrac{19}{12}$;
   B. $\dfrac{4}{5}$;
   C. $\dfrac{8}{12}$;
   D. $\dfrac{15}{24}$;
   E. $\dfrac{9}{10}$.

   Answer: A

   Solution: The lowest common denominator between 4 and 6 is 12. So, basically, we need to rewrite the fraction $\dfrac{3}{4}$ with denominator 12 instead of 4. So, we need to multiply both bottom and top by 3. A similar operation happens with $\dfrac{5}{6}$. Answer is $\dfrac{(3)(3)}{(4)(3)} + \dfrac{(5)(2)}{(6)(2)} = \dfrac{9}{12} + \dfrac{10}{12} = \dfrac{19}{12}$.

3. (Competition 2011) Evaluate $\dfrac{\frac{4}{5}}{\frac{5}{8}}$:

   A. $\dfrac{1}{2}$;
   B. 2;
   C. $\dfrac{9}{13}$;
   D. $\dfrac{32}{25}$;

E. $\frac{52}{40}$.

Answer: D

Solution: Division of fractions: keep the top fraction and multiply by the inverse of the bottom fraction. Answer is $\left(\frac{4}{5}\right)\left(\frac{8}{5}\right) = \frac{32}{25}$.

4. (Practice 2012) What is the number that is one half of one quarter of one tenth of 400?

   Answer: 5

   Solution: Most of the time, the word "of" will translate into multiplication. Number is $\left(\frac{1}{2}\right)\left(\frac{1}{4}\right)\left(\frac{1}{10}\right)(400) = 5$.

5. (Competition 2011) If $x = -2$ and $y = 5$, what is the value of the expression $2x^3 - 3xy$?

   A. 14
   B. 46
   C. 54
   D. -46
   E. -54

   Answer: A

   Solution: It is simply substitution, as a hint, we need substitute with parentheses to make sure we do not lose any negative sign. Answer is $2(-2)^3 - 3(-2)(5) = 2(-8) + 6(5) = -16 + 60 = 14$.

6. (Competition 2016) What is 40% of 190?

   Answer: 76

   Solution 1: Calculating 10% is easy, it is simply dividing the number by 10. After that, we use that 40% is four times 10%. Since 10% of 190 is 19. So, 40% will be $4 \times 19 = 76$.

   Solution 2: We "translate" the expression into math, using that "$\% = \frac{1}{100}$" and "of is the same as times": 40% of $190 = 40\left(\frac{1}{100}\right)(190) = \left(\frac{4}{10}\right)(190) = 76$.

7. (Competition 2011) In a class of 25 students, 11 studied French. What percent studied French?

   A. 25%

B. 11%
C. 44%
D. Approximately 2.72%

Answer: C

Solution: To find percentage, we need to write the fraction (proportion) as a fraction having 100 as denominator. Answer is $\frac{11}{25} = \frac{(11)(4)}{(25)(4)} = \frac{44}{100} = 44\%$.

8. (Competition 2011) Bob's annual salary was $24,000 last year. This year he received a 4% raise. What is his annual salary this year?

    A. $24,960
    B. $25,040
    C. $26,120
    D. $30,000

Answer: B

Solution 1: The raise was $\frac{4}{100} 24{,}000 = 4(240) = 960$. Hence, Bob's new salary is $24,960.

Solution 2: Bob's new salary is 104% of $24,000: $\frac{104}{100} 24000 = 104(240) = 24{,}960$.

9. (Practice 2012) A Spanish teacher has noticed that about 35% of those taking the test make an A, historically. If 135 students take the test, how many can be expected to make an A?

Answer: 47 or 47.25 (both answers were accepted as correct)

Solution: It is expected to have again 35% of the total students getting A. So, $\frac{35}{100}(135) = 47.25$.

10. (Competition 2015) If you raise 2% the smaller of two consecutive positive integers, the result is the largest. What is the largest number? (Enter DNE if there is no enough info to solve the problem)

Answer: 51

Solution 1: The numbers are x and $x+1$. When we raise a number 2%, we are considering 102% of that number. We need to solve $\frac{102}{100}x = x+1$. So, $102x = 100x + 100$. Then $x = 50$. So the largest is 51.

Solution 2: It is similar to solution 1, basically we need to find a number whose 2%, or $1/50$, is equal to 1: $\frac{1}{50}x = 1$. The lowest number is 50, so the largest will be 51.

11. (Competition 2011) Jane donates 2% of her annual salary to charity and Sam donates 1% of his annual salary to charity. The total of Jane and Sam's donations is $620. If Jane makes $1000 more than Sam, how much is Jane's annual salary?

    A. $20,000
    B. $21,000
    C. $24,000
    D. $30,000

Answer: B

Solution: Call j Jane's salary and s Sam's salary. Equations representing the situation are:

$0.02j + 0.01s = 620$

$j = s + 1000$

Substituting, we get $0.02(s + 1000) + 0.01s = 620$. So, $0.03s + 20 = 620$. Then, $s = \frac{600}{0.03} = 20,000$, hence $j = 21,000$.

12. (Competition 2011) Find all the solutions for $x^2 - 4x - 45 = 0$

    A. $x = 9$ and $x = 5$;
    B. $x = 9$ and $x = -5$;
    C. $x = -9$ and $x = 5$;
    D. $x = -9$ and $x = -5$;
    E. None of above

Answer: B

Solution 1: Using quadratic formula $x = \frac{-b \pm \sqrt{b^2 - 4ac}}{2a} = \frac{-(-4) \pm \sqrt{(-4)^2 - 4(1)(-45)}}{2(1)} = \frac{4 \pm \sqrt{16 + 180}}{2} = \frac{4 \pm \sqrt{196}}{2} = \frac{4 \pm 14}{2}$. Lowest root is $\frac{4 - 14}{2} = -5$, the other is $\frac{4 + 14}{2} = 9$.

Solution 2: Factoring, $(x + 5)(x - 9) = 0$. So, either $x + 5 = 0$ or $x - 9 = 0$. Hence, $x = -5$ or $x = 9$.

Solution 3: We know that the sum of the roots of a second-degree polynomial is $Sum = -\frac{b}{a}$, while the product is $Product = \frac{c}{a}$. Then, we need to look for two numbers such that their product is -45 (so they have different signs), and sum is $+4$ (so the positive root has a bigger absolute value).

Solution 4: There are four different values as "candidates": $-9, -5, 5, and\ 9$. Let's try each

$x =$	$x^2 - 4x - 45 =$	Root?
-9	$(-9)^2 - 4(-9) - 45 = 81 + 36 - 45 = 72$	No.
-5	$(-5)^2 - 4(-5) - 45 = 25 + 20 - 45 = 0$	Yes.
5	$(5)^2 - 4(5) - 45 = 25 - 20 - 45 = -40$	No.
9	$(9)^2 - 4(9) - 45 = 81 - 36 - 45 = 0$	Yes.

13. (Competition 2011) Find all the solutions for $x^2 + 4x + 45 = 0$.

A. $x = 9$ and $x = 5$;
B. $x = 9$ and $x = -5$;
C. $x = -9$ and $x = 5$;
D. $x = -9$ and $x = -5$;
E. None of above.

Answer: E

Solution 1: Using quadratic formula
$x = \frac{-b \pm \sqrt{b^2 - 4ac}}{2a} = \frac{-(4) \pm \sqrt{(4)^2 - 4(1)(45)}}{2(1)} = \frac{-4 \pm \sqrt{16 - 180}}{2} = \frac{4 \pm \sqrt{-164}}{2}$. So, the solutions are not real.

Solution 2: There are four different values as "candidates": $-9, -5, 5, and\ 9$. Let's try each

$x =$	$x^2 + 4x + 45 =$	Root?
-9	$(-9)^2 + 4(-9) + 45 = 81 - 36 + 45 = 90$	No.
-5	$(-5)^2 + 4(-5) + 45 = 25 - 20 + 45 = 50$	No.
5	$(5)^2 + 4(5) + 45 = 25 + 20 + 45 = 90$	No.
9	$(9)^2 + 4(9) + 45 = 81 + 36 + 45 = 162$	No.

14. (Competition 2016) When simplified $\left(-\frac{1}{125}\right)^{-2/3}$ becomes:

A. $\frac{1}{25}$ B. $-\frac{1}{25}$ C. 25 D. -25 E. $25\sqrt{-1}$ F. None of previous

Answer: C

Solution: This was to test negative numbers as exponents in which we have to flip the base:
$\left(-\dfrac{1}{125}\right)^{-2/3} = \left(-\dfrac{125}{1}\right)^{2/3} = (\sqrt[3]{-125})^2 = (-5)^2 = 25$.

15. (Competition 2015) The result of the division of $4^{(4^2)}$ by 4^4 is a number that can be written as 4^x. What is the value of x?

    Answer: 12

    Solution: We use a very basic rule for dividing powers with same base: $\dfrac{a^b}{a^c} = a^{b-c}$. One number is 4^{16} the second is 4^4. So, $\dfrac{4^{16}}{4^4} = 4^{12}$.

16. (Competition 2016) If the average of x and 9 is 7, what is the value of x?

    Answer: 5

    Solution: Average between two numbers is the result of adding them and dividing by 2. Hence: $\dfrac{x+9}{2} = 7$, then $x + 9 = 14$, so $x = 5$.

17. (Practice 2014) Subtracting the same number from both numerator and denominator of $\dfrac{13}{14}$ we get the fraction $\dfrac{14}{13}$. What is such number?

    Answer: 27

    Solution: Call x the number, so $\dfrac{13-x}{14-x} = \dfrac{14}{13}$. Hence, $13(13-x) = 14(14-x)$. So, $169 - 13x = 196 - 14x$. Finally, $x = 27$. In fact, $\dfrac{13-27}{14-27} = \dfrac{-14}{-13}$.

18. (Competition 2014) If all the stars represent the same number in the expression

    $$\dfrac{*}{*} - \dfrac{*}{6} = \dfrac{*}{12}$$

    What is the value of *?

    Answer: 4.

Solution: To make it more familiar, instead of a star (or asterisk) we write x. Then, $\frac{x}{x} - \frac{x}{6} = \frac{x}{12}$. Multiplying both sides by 12: $12 - 2x = x$. Hence, $x = 4$.

19. (Competition 2015) When travelling by aircraft, passengers have a maximum allowable weight for their luggage. They are then charged £10 for every kilogram overweight. If a passenger carrying 40 kg of luggage is charged £50, how much would a passenger carrying 80 kg be charged?

Answer: 450

Solution 1: Since each kilogram costs £10, a charge of £50 means the passenger has 5 kilograms in overweight. Hence, the maximum allowed weight is 35 kilograms. A passenger with 80 kg, will have $80 - 35 = 45$ kg in overweight, so he'd pay £450.

20. (Competition 2011) Michelle's cellular-phone company offers a plan that allows 300 minutes of use for $29.95 per month and charges $0.20 for each additional minute. All prices include tax and fees. Michelle has budgeted $50 per month for calls on her cellular phone. What is the maximum number of minutes that she can use her cellular phone each month without spending more than $50?

  A. 101 min
  B. 100 min
  C. 401 min
  D. 400 min

Answer: D

Solution 1: That's the solution of an inequality: if x is the amount of extra minutes (after 300 minutes) then her monthly cost is $C(x) = 29.95 + 0.20x$, assuming x is a positive integer. So, $29.95 + 0.20x \leq 50$. Then $0.20x \leq 20.05$, so $x \leq 100.25$. The maximum x would be 100 extra minutes, which allows Michelle to talk for 400 minutes, since 300 minutes are already included in the plan.

Solution 2: It is pretty much the same solution, but with descriptions instead of equations. Her budget allows a maximum of $50, since the plan is $29.95, Michelle has $20.05 for extra minutes. 100 extra minutes would cost $(100)(0.20) = 20$ dollars, while 101 extra minutes would cost $20.20, which is more than what she has.

21. (Practice 2013) To exercise on Lake Texana Park, Morgan bought a mountain bike for $750. He put 20% of the cost as an initial down payment, and paid off the remainder of the cost in 10 equal monthly payments, with no interest. What is the price (in $) of one monthly payment?

Answer: 60

Solution: Morgan paid 20% of $750, which is $150. The $600 left was divided into 10 payments of $\frac{600}{10} = 60$ dollars.

22. (Competition 2015) Ricardo computed the average of two different positive integers and the result was 98. Both numbers are 2-digit. What is the difference between the larger and the smaller? (Enter DNE if there is no enough info to solve the problem)

Answer: 2

Solution: The integers are x and y, with both numbers being less than 100, but more than 9. Since they are different numbers, they can't be both 98, one needs to be more than 98 other needs to be less. The only possibility if the pair 97 and 99. So, the difference between them is 2.

23. (Competition 2011) Wesley and Delia are playing a math game. Wesley gives Delia these steps to follow.

Step 1: Multiply a number by 6 and then subtract 4.

Step 2: Divide the result by 2.

Step 3: Add 3 to the result from the second step.

If Delia's final answer is 19, what was the original number?

A. 10/3
B. 12
C. 8
D. 6

Answer: D

Solution 1: We can work the problem backwards. After step 3, Delia's number is 19 so that implies that before step 3, her number was 16. This means that before step 2, her number was 32. Finally, if we undo the "subtract 4" part, we know that after "multiply by 6", Delia had 36. So, she started with 6.

Solution 2: We can use "composition of functions". First step can be represented by the function $f(x) = 6x - 4$. The function for second step is $g(x) = \dfrac{x}{2}$. And $h(x) = x + 3$ symbolizes last step. All steps, in order will be: $h(g(f(x)))$. Let's go step-by-step: $g(f(x)) = \dfrac{6x-4}{2} = 3x - 2$, so $h(g(f(x))) = (3x - 2) + 3$. We need to solve: $(3x - 2) + 3 = 19$. So, $3x + 1 = 19$. Then $3x = 18$. Hence, $x = 6$.

24. (Competition 2014) The figure below represents a map of roads from Rio de Janeiro to Sao Paolo. The numbers tell how much each toll road costs, in Reais (Brazilian currency). What is the minimum that can be spent in tolls from Rio de Janeiro to Sao Paolo?

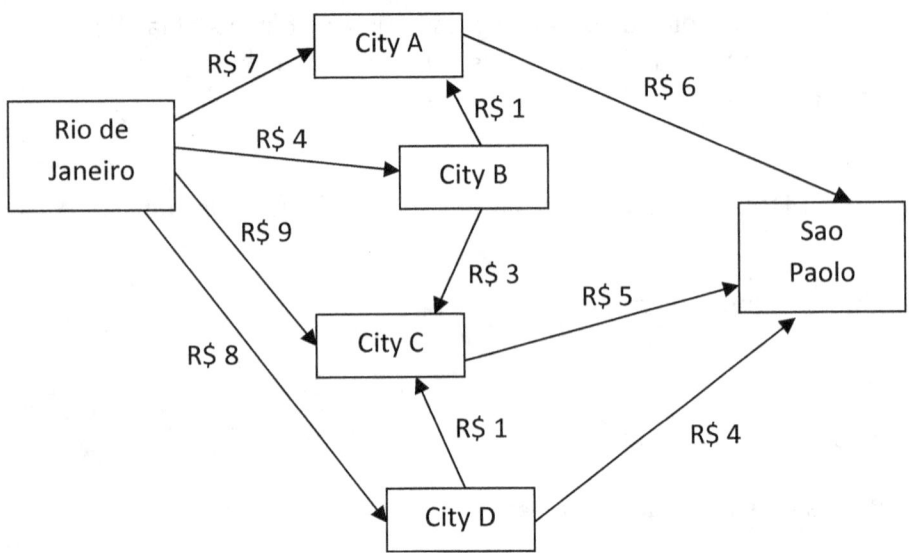

Answer: $11

Solution: A visual path can be constructed, but for completion we make a broader analysis. From city A to Sao Paolo, toll road is $6. From city B, minimum is $7 (there are two possibilities). From city C, toll is $5. From city D, minimum is $4 (two options). From Rio de Janeiro to city A, toll is $7, making the total toll for Rio-A-SP equal to $13. From Rio to city B, toll is $4, and Rio-B-SP is $11. Rio to city C costs $9, so Rio-C-SP costs $14. Finally, Rio to city D costs $8, so Rio-D-SP costs $12. Minimum is, then, $11.

25. (Practice 2012) Justin purchased $30,000 worth of municipal bonds. If he can purchase them in denominations of $500 or $1,000, what is the maximum number of bonds he purchased?

Answer: 60

Solution: The maximum number of bonds would be when all his bonds are $500, so 60 bonds.

26. (Practice 2012) Esther has a bucket with 9 liters of a mixture that contains 50% alcohol and 50% water. She wants to add water to the mixture so that only 30% of the solution is alcohol. How many liters of water will she add?

Answer: 6

Solution: Esther's mixture has 4.5 liters of water and 4.5 liters of alcohol. If she adds x liters of water, then the mixture would have $4.5 + x$ liters of water, and a total of $9 + x$ liters. We want

$4.5 = 30\% \, (9+x) = \dfrac{30}{100}(9+x) = \dfrac{3}{10}(9+x)$. Multiplying by 10: $45 = 27 + 3x$. So, $3x = 18$, hence $x = 6$.

28. **(Competition 2014)** Andy went shopping. In the first store, he spent half of the money plus 10 dollars. In the second store, he spent half of he had left plus 5. In the third and final store, he spent one third of what he had left plus 3. He finished his day with $11. With how much did he start the day?

    Answer:

    Solution: We should work the problem backwards. At the last stop, Andy had x dollars when he arrived, he spent $\dfrac{x}{3} + 3$, leaving him with $x - \left(\dfrac{x}{3} + 3\right)$. So, $\dfrac{2x}{3} - 3 = 11$. Solving the equation, we calculate that he had $21 when he arrived there.

    On the second store, he got there with y dollars, and spent $\dfrac{y}{2} + 5$, leaving him with $y - \left(\dfrac{y}{2} + 5\right) = \dfrac{y}{2} - 5$. Since this quantity must be equal to the amount he entered the third store: $\dfrac{y}{2} - 5 = 21$, so $y = 52$.

    Finally, he started his day with z dollars, spent $\dfrac{z}{2} + 10$. Then $\dfrac{z}{2} - 10 = 52$. He started the day with $124.

28. **(Competition 2014)** Each amoeba divides into two separate amoebas once every hour. If we start with only one amoeba, how many will there be at the end of eight hours? (Mathematical amoebas never die!)

    Answer: 256

    Solution: Start with one amoeba. At the end of the first hour, there are $2 = 2^1$. Then, at the end of the second hour, there are $4 = 2^2$, and so on. At the end of the eighth hour, there would be $2^8 = 256$ amoebas.

29. **(Competition 2015)** Suppose x and y are different numbers satisfying $x - \dfrac{1}{x} = y - \dfrac{1}{y}$. What is the value of the product xy? (Enter DNE if there is no enough info to solve the problem)

    Answer: -1

    Solution: First of all, neither x nor y can be zero, since they are in the denominator. We take the common denominator which is xy. $\dfrac{x^2y - y}{xy} = \dfrac{xy^2 - x}{xy}$. Then, $x^2y - y = xy^2 - x$. So, $x^2y + x = xy^2 + y$. Then, $x(xy + 1) = y(xy + 1)$. Then, there are two possibilities: either we can cancel $(xy + 1)$ on both sides, meaning that the expression is not zero, or we can't cancel, since the expression is zero. If we assume that the expression is not zero, then we would get $x = y$, but the question says that x and y are different. Hence the only possibility is $xy + 1 = 0$. So, $xy = -1$.

30. (Competition 2015) Alex starts to go down on a staircase, having 24 steps, at the same time that Erin started to go up. Both of them are going at constant (different) speeds. Alex had gone down ¾ of the length when she crossed with Erin. At the moment that Alex finished going down, how many steps is left for Erin to go up?

    Answer: 16

    Solution: Since Alex met Erin when she had ¾ of the steps, they met on step 6, counting from bottom up. Alex needed one third of the time used to complete her way. That is enough time for Erin go up one third more of what she accomplished. So, Erin would go to step 8. Hence, there are still 16 steps left for Erin.

31. (Competition 2012, and Competition 2013 with slightly different numbers) Kathryn must make at least 92 average on her tests to make an A in her math class, and get a new vehicle. If she has 88, 95, 83, and 97 on her previous tests, what would the smallest grade that she need to make on her last test to get the A on her class?

    Answer: 97

    Solution 1: Let x be the grade she gets on the last test, her average would be $\frac{88 + 95 + 83 + 97 + x}{5} = \frac{363 + x}{5}$. We need this to be at least 92: $\frac{363 + x}{5} \geq 92$. Then $363 + x \geq 460$. So, $x \geq 97$.

    Solution 2: We make an analysis of the difference from the grade and 92. So, on the first exam, she was 4 points below 92. On the second, she was 3 points above. On the third, she got 9 points below. And on the fourth, she got 5 points above. A total of $(-4) + 3 + (-9) + 5 = -5$. To compensate, she needs to get 5 points above 92.

32. (Competition 2012) On a multiple choice test with 24 problems, each problem receives the following scores: four if the answer is correct, negative one if it is incorrect or zero if left blank. Robert received 52 points on the test, what is the maximum number of correct answers that he received.

    Answer: 15

    Solution: Let c be the amount of correct answers, w be the amount of wrong answers and b be the amount of questions left blank. So, $c + w + b = 24$. The amount of points is $4c - w = 52$. So, $4c = 52 + w$. In other words, $52 + w$ must be a number greater than or equal to 52 and divisible by 4. Since $w \leq 24$, the candidates for $4c$ are 52 (if $w = 0$), 56 (if $w = 4$), 60 (if $w = 8$), 64 (if $w = 12$), 68 (if $w = 16$), 72 (if $w = 20$) and 76 (if $w = 24$). However, not every possibility is valid, since $c + w \leq 24$. Let's check: $w = 0 \Rightarrow c = 13$, $w = 4 \Rightarrow c = 14$, $w = 8 \Rightarrow c = 15$, as if $w \geq 9 \Rightarrow c > 15$, we see that the maximum number of correct answers is 15.

33. (Practice 2012) Four years ago, Jane was twice as old as Sam. Four years on from now, Sam will be 3/4 of Jane's age. How old is Jane now?

Answer: 12

Solution 1: Call Jane's age today j, and Sam's s. Four years ago Jane was $j-4$ and Sam was $s-4$. The first piece of info can be written as $j-4=2(s-4)$. Four years from now, Jane will be $j+4$, and Sam will be $s+4$. So, the second equation is $s+4=\frac{3}{4}(j+4)$. The first equation becomes $j+4=2s$, which can be used on the second equation to become: $s+4=\frac{3}{4}(2s)=\frac{3}{2}s$. Then s

Solution 2: Another approach would be by cleverly guessing. Jane was older than Sam, so she still is. Also, Sam and Jane need to be at least more than 4 years old. Also, four years from now, Jane needs to have an age that is divisible by 4, since Sam's age would be $3/4$ of that. Hence, Jane's age is divisible by four now as well. Jane could be 8, 12, 16, 20, ... If Jane is 8 now, she was 4 four years ago and, from the first piece of info, Sam was 2. But four years from now, Jane would be 12 and Sam 10, and $10 \neq \frac{3}{4}(12)$. Because the first equation is fulfilled and the second is not, but the difference is small, we try the next possible number, $j=12$, which is our answer.

34. (Competition 2013) On a multiple choice test with 24 problems, each problem receives the following scores: four if the answer is correct, negative one if it is incorrect or zero if left blank. Robert received 52 points on the test, what is the maximum number of correct answers that he received?

Answer: 15

Solution: Let c be the amount of correct answers, w be the amount of wrong answers and b be the amount of questions left blank. So, $c+w+b=24$. The amount of points is $4c-w=52$. So, $4c=52+w$. In other words, $52+w$ must be a number greater than or equal to 52 and divisible by 4. Since $w \leq 24$, the candidates for $4c$ are 52 (if $w=0$), 56 (if $w=4$), 60 (if $w=8$), 64 (if $w=12$), 68 (if $w=16$), 72 (if $w=20$) and 76 (if $w=24$). However, not every possibility is valid, since $c+w \leq 24$. Let's check: $w=0 \Rightarrow c=13$, $w=4 \Rightarrow c=14$, $w=8 \Rightarrow c=15$, as if $w \geq 9 \Rightarrow c>15$, we see that the maximum number of correct answers is 15.

35. (Competition 2015) At a birthday party, one-half drank only lemonade, one-third drank only cola, fifteen people drank neither, and nobody drinks both. How many people were there at the party?

Answer: 90

Solution: Call l the number of people that drank lemonade, and c the number of people who drank cola. The total number of people at the party was $l+c+15$. We were told that l is half of this amount, while c is a third:

17

$l = \frac{1}{2}(l + c + 15)$, multiplying by 2: $2l = l + c + 15$. Solving: $l = c + 15$. $c = \frac{1}{3}(l + c + 15)$, multiplying by 3: $3c = l + c + 15$. Solving: $2c = l + 15$.

Hence, $2c = (c + 15) + 15$. So, $c = 30$. Then, $l = 45$. And there were 90 people on the party.

36. (Competition 2013) At Victoria Zoo, there are bulls, gooses and snakes. There are 10 heads total, and 22 feet. Also, we know that there are fewer gooses than bulls. How many gooses are there?

    Answer: 1 or 3 (half credit was given if a student mark only one of the possible numbers)

    Solution: Let b be the number of bulls, g the number of gooses and s the number of snakes. We have 10 heads: $b + g + s = 10$. We have 22 feet: $4b + 2g = 22$, or $2b + g = 11$ (hence g needs to be odd, so b is an integer). We also know that $g < b$. Let's see the possibilities:

g	b	s
1	5	4
3	4	3
5	3	2
7	2	1
9	1	0

    The last three options does not satisfy our problem, since we need $g < b$. The first and second options are valid, hence the answer is supposed to be "1 or 3".

37. (Practice 2012) Look at the drawing. The numbers alongside each column and row are the total of the values of the symbols within each column and row. What should replace the question mark?

    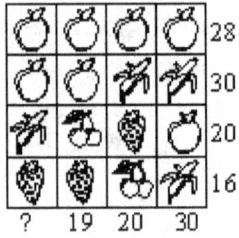

    Answer: 25

    Solution: Look at the first row: 4 apples is 28, so each apple is 7. Now go to the second row, 2 apples (14) plus 2 corns equal 30, so 2 corns is 16, making each corn 8. On third and fourth rows, there are strawberries and cherries, we could do a system of equations and solve, call s for strawberry and c for cherry:

    $8 + c + s + 7 = 20$
    $2s + c + 8 = 16$

Hence $c + s = 5$ and $2s + c = 8$. We find that $s = 3$ and $c = 2$.

Now the question mark is the sum of 2 apples, 1 corn, and 1 strawberry: $2(7) + 8 + 3 = 25$

38. (Competition 2014) Ricardo is twice as old as Jeffrey was when Ricardo was half as old as Jeffrey will be when Jeffrey is three times as old as Ricardo was when Ricardo was three times as old as Jeffrey. Their combined age makes forty-eight. How old is Ricardo now?

Answer: 30

Solution: Let x be Ricardo's present age and y be Jeffrey's present age.

i. Their combined age is 48: $x + y = 48$.
ii. Ricardo now (age x) is twice as old as Jeffrey was z years ago: $x = 2(y - z)$
iii. Ricardo was (z years ago) half as old as Jeffrey will be w years from now: $x - z = \frac{1}{2}(y + w)$
iv. Jeffrey will be (w years from now) three times as old as Ricardo was v years ago: $y + w = 3(x - v)$
v. Ricardo was (v years ago) three times as old as Jeffrey was (v years ago): $x - v = 3(y - v)$.

There are five unknowns, x, y, z, v, w, and five equations. That should be enough to find out each variable.

$x + y = 48$
$x - 2y + 2z = 0$
$2x - y - 2z - w = 0$
$3x - y - 3v - w = 0$
$x - 3y + 2v = 0$

There are several methods to solve, but let's try a simple substitution method. From the first equation: $y = 48 - x$. Then (suggestion, take a piece of paper and follow our equations from 2nd to 5th):

$3x + 2z = 96$
$3x - 2z - w = 48$
$4x - 3v - w = 48$
$2x + v = 72$

We still have 4 equations with 4 unknowns. It seems that the last one may be simpler to substitute: $v = 72 - 2x$. Equations 2, 3, and 4 can be rewritten as:

$3x + 2z = 96$
$3x - 2z - w = 48$
$10x - w = 264$

Then, $w = 10x - 264$:

$3x + 2z = 96$
$7x + 2z = 216$

Now, we can subtract equations:

$4x = 120$

So $x = 30$. For completion: $y = 18, z = 3, w = 36, v = 12$.

39. (Practice 2012) If two typists can type two pages in two minutes, how many typists will it take to type 18 pages in six minutes?

    Answer: 6

    Solution 1: Notice that if there are more typists, more pages can be typed on the same amount of time (fixing the number of minutes). So, number of typists is directly proportional to the number of pages. If we compare number of typists and number of minutes to do a specific job (fixing the number of pages): more typists imply less minutes. So, number of typists is inversely proportional to number of minutes.

Number of Typists	Number of pages	Number of minutes
2 ↑	2 ↑	2 ↓
x	18	6

    $$\frac{2}{x} = \left(\frac{2}{18}\right)\left(\frac{6}{2}\right)$$

    $$\frac{2}{x} = \frac{1}{3}$$

    $$x = 6$$

    Solution 2: We can imagine that, on the first situation, each typist types a page, and it takes two minutes. So, in 6 minutes, each typist can type 3 pages. Since, there are 18 pages to be typed, we would need 6 typists.

40. (Competition 2016) Gisele could paint her entire house in 40 days. Her brother is faster, and could complete the task in 24 days. If they work together, how many days would they need to paint Gisele's house?

    Answer: 15

    Solution: Let's solve this by "productivity". Gisele completes $\frac{1}{40}$ of the service per day. Her brother completes $\frac{1}{24}$ of the service per day. Together, they complete $\frac{1}{40} + \frac{1}{24} = \frac{3+5}{120} = \frac{8}{120} = \frac{1}{15}$ per day. So, they need 15 days.

41. (Competition 2015) Matilda's father takes 20 minutes to mow the back garden lawn, and Matilda takes 30 minutes to do the same job. If they worked together, how long (in minutes) would it take to cut the lawn?

    Answer: 12

Solution: Matilda's father can do $\frac{1}{20}$ of the job per minute, while she can do only $\frac{1}{30}$ per minute. Together, they can do $\frac{1}{20} + \frac{1}{30} = \frac{5}{60} = \frac{1}{12}$ of the job per minute. Hence, together, they'd need 12 minutes.

42. (Competition 2012, and Competition 2014) What is the third non-zero digit on the decimal representation of the fraction $\frac{1}{5^8}$? (Hint: There is a faster way other than making the computation)

    Answer: 6

    Solution: The question becomes trivial is we make a small computation:
    $\frac{1}{5^8} = \frac{1}{5^8} \cdot \frac{2^8}{2^8} = \frac{2^8}{10^8} = \frac{256}{10^8}$. Hence, the decimal representation of the number will have "6" as the third non-zero digit.

43. (Practice 2012) If it were two hours later, it would be half as long until midnight as it would be if it were an hour later. What time is it now?

    Answer: 9

    Solution: Let t be the time now, an hour later will be $t + 1$, two hours later will be $t + 2$. Hours until midnight is computed as $(12 - time)$. So, the equation is $12 - (t + 2) = \frac{1}{2}[12 - (t + 1)]$. Multiplying by 2, we have $24 - 2t - 4 = 12 - t - 1$. So, $t = 9$.

44. (Practice 2012 and Competition 2013) Let a and b be non-zero numbers such that the equation $x^2 + ax + b = 0$ has a and b as roots. Find the value of a.

    Answer: 1

    Solution: A second-degree equation with the quadratic term simply x^2 can be rewritten as $x^2 - Sx + P = 0$, where S is the sum of its roots, and P is the product. In this case, $S = a + b = -a$, and $P = ab = b$. From the last equation, since $a \neq 0$, $a = 1$. Then $1 + b = -1$, so $b = -2$.

45. (Competition 2015) My father and I share the same birthday. When I was 14 years old my father was 42 years old, which was three times my age. Now he is twice my age, how old am I?

    Answer: 28

    Solution: The difference between the ages remain the same: $42 - 14 = 28$. After x years, the person will be $14 + x$, his father will be $42 + x$. The equation is: $42 + x = 2(14 + x) = 28 + 2x$. Hence $x = 14$.

So, his age now is $14 + 14 = 28$.

46. (Competition 2015) Two railway stations, P and Q, are 279 miles apart. A train departs from P at 2pm and travels at a constant speed of 51 mph towards Q. At 3pm a second train begins a journey from Q towards P at a constant speed of 60 mph. How far apart (in miles) are the two trains twenty minutes before they pass each other?

Answer: $\dfrac{111}{3}$

Solution: Although the question may seem hard the first time you read it, it is quite simple. The relative speed between the trains is $51 + 60 = 111$ mph. So, 20 minutes (or $1/3$ of an hour) before they pass each other, the trains will be $\dfrac{111}{3}$ miles apart.

47. (Competition 2016) There are five positive integers: $a \leq b \leq c \leq d \leq e$. Number 1 is the only number that is repeated. The quantity of numbers to the left of the number 5 is the same as to its right. The average of the numbers is 4. What is the biggest number?

Answer: 7

Solution: This was originally a question involving the statistical terms "mode", "mean" and "median". Since the quantity of numbers to the left of 5 is the same as to the right (in other words, 5 is the median), then $c = 5$. Since 1 is the only number that is repeated (making 1 the mode), $a = b = 1$. The average of the numbers (i.e., the mean) is 4, so $\dfrac{1 + 1 + 5 + d + e}{5} = 4$. Then $d + e = 13$. The only possibility is $d = 6$ and $e = 7$.

48. (Practice 2012) What is the value of the expression $20132013^2 + 20132005^2 - 32$

    A. 2×20132005^2
    B. 2×20132013^2
    C. 2×20132009^2
    D. 2×20132005
    E. 2×20132009

Answer: C

Solution: Notice that $20132013 = 20132009 + 4$, while $20132005 = 20132009 - 4$. Using that $(a \pm b)^2 = a^2 \pm 2ab + b^2$.

We rewrite the expression as

$(20132009 + 4)^2 + (20132009 - 4)^2 - 32 =$

$20132009^2 + 2 \times 20132009 \times 4 + 4^2 + 20132009^2 - 2 \times 20132009 \times 4 + 4^2 - 32 = 2 \times 20132009^2$.

Chapter 2: Graphs

Reading graphs and getting basic graphs on the plane

We also included some basic questions regarding how to interpret graphical data, and some questions on lines and parabolas.

For graphical data, the hint is to read and understand each graph and each description. There might be some details that need attention.

For lines, one approach is that non-vertical lines can be expressed as $y = mx + b$, where m is the slope, and b is the y-intercept. Slope is given as "rise over run" or "change in y divided by change in x".

49. (Competition 2011) Scientists have stocked Riverside's pond with a species of fish. The scientists noted that the population has steadily decreased over a period of time until the population is approximately half the number of fish originally stocked. If the number of fish are plotted on the y-axis and the amount of time on the x-axis, which of the following could result?

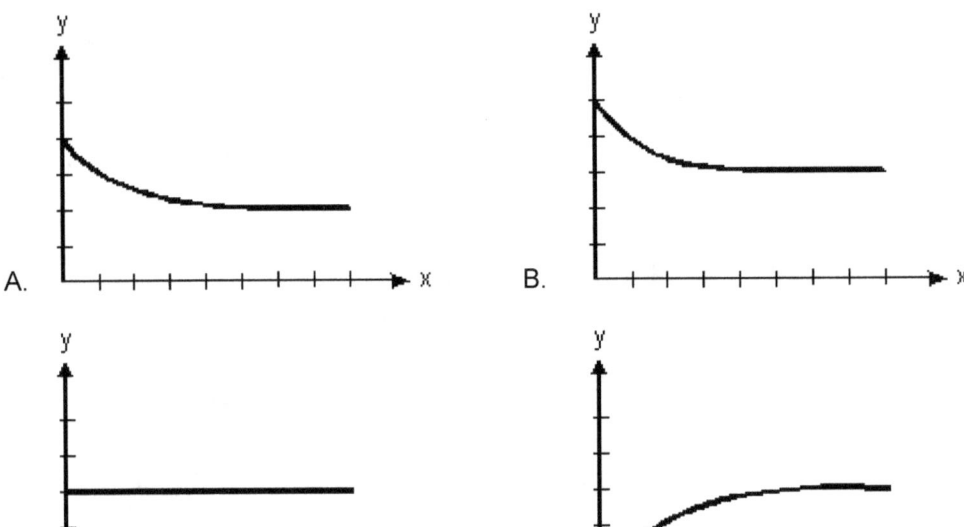

Answer: A

Solution: On graph C, the y-quantity does not vary with respect to the x-quantity. Graph D shows y-quantity increasing when x-quantity increases. Graphs A and B are decreasing. Since the population of fish (y-quantity) is decreasing with respect to time (x-quantity), the graph needs to be decreasing. On graph A, the initial population is at the 4th mark, decreasing to around the 2nd, while on graph B population decreases from 5th to 3rd mark.

50. (Practice 2011) Take a look at the bar graph below. It shows the introductory courses at a random university and their enrolment this past semester.

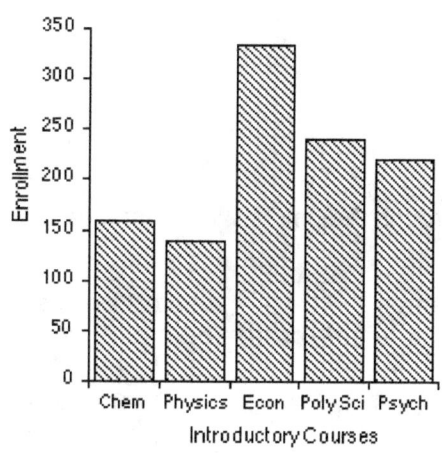

Now answer: according to the graph above, which course has the most students enrolled in it?

   A. Chemistry
   B. Physics
   C. Economics
   D. Political Sciences
   E. Psychology

Answer: C

Solution: It is the highest bar.

51. (Competition 2011) Use the graph below to answer the question that follows.

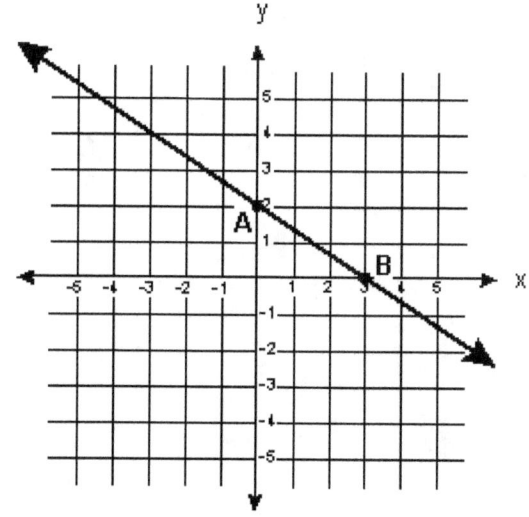

Which of the following equations represents line *AB*?

   A. $y = -2x/3 + 2$

25

B. $y = 3x/2 + 3$
C. $y = -2x + 3$
D. $y = 3x + 2$

Answer: A

Solution 1: Line passes through the points $(0,2)$ (that means $x = 0, y = 2$) and $(3,0)$. We use that non-vertical lines can be written as $y = mx + b$, where m is the slope, given by the formula $m = \dfrac{y_2 - y_1}{x_2 - x_1}$ (change in y divided by change in x, or "rise over run"). And b is the y-intercept (y-coordinate on the point where the line crosses the y-axis). So, $m = \dfrac{0 - 2}{3 - 0} = -\dfrac{2}{3}$, and $b = 2$. So, $y = -\dfrac{2}{3}x + 2$.

Solution 2: Alternatively, we can simply plug the values of each point and see which equations would be satisfied. If we plug $x = 0$:

	$x = 0$	$x = 3$
A. $y = -\dfrac{2x}{3} + 2$	$y = 2$, correct	$y = 0$, correct
B. $y = \dfrac{3x}{2} + 3$	$y = 3$, incorrect	$y = \dfrac{15}{2}$, incorrect
C. $y = -2x + 3$	$y = 3$, incorrect	$y = -3$, incorrect
D. $y = 3x + 2$	$y = 2$, correct	$y = 8$, incorrect

52. (Practice 2011) Use the graph below to answer the question that follows.

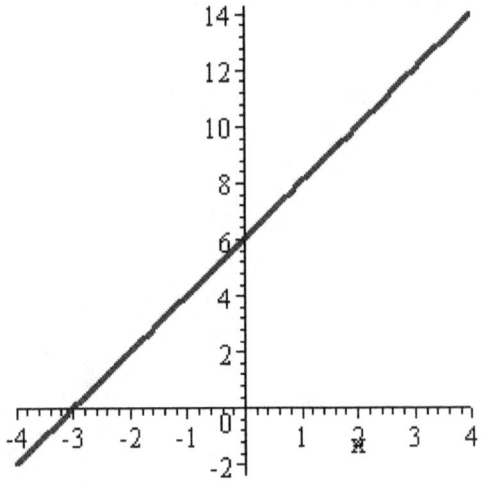

Which of the following equations represents line above?

A. $y = 6x - 3$
B. $y = -3x + 6$
C. $y = 3x + 6$
D. $y = 2x + 6$

Answer: D

Solution 1: Line passes through the points $(-3, 0)$ and $(0, 6)$. We use that non-vertical lines can be written as $y = mx + b$, where $m = \dfrac{y_2 - y_1}{x_2 - x_1}$ is the slope. And b is the y-intercept. So, $m = \dfrac{6 - 0}{0 - (-3)} = \dfrac{6}{3} = 2$, and $b = 6$. So, $y = 2x + 6$.

Solution 2: Alternatively, we can simply plug the values of each point and see which equations would be satisfied. If we plug $x = 0$:

		$x = -3$	$x = 0$
A.	$y = 6x - 3$	$y = -21$, incorrect	$y = -3$, incorrect
B.	$y = -3x + 6$	$y = 15$, incorrect	$y = 6$, correct
C.	$y = 3x + 6$	$y = -3$, incorrect	$y = 6$, correct
D.	$y = 2x + 6$	$y = 0$, correct	$y = 6$, correct

53. (Practice 2011) Find the intersection between the two lines:

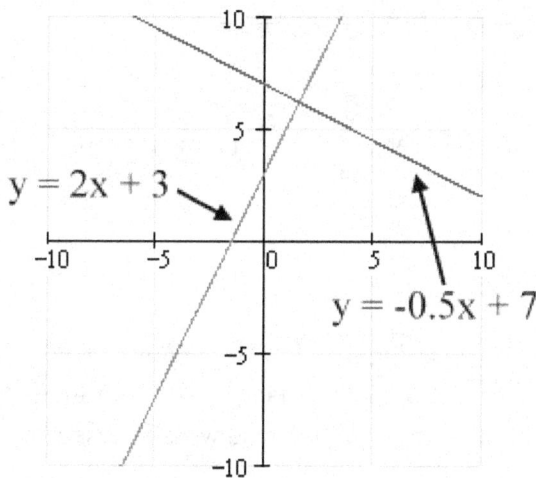

A. $\left(\dfrac{31}{5}, \dfrac{8}{5}\right)$

B. $\left(\dfrac{3}{2}, 6\right)$

C. $\left(\dfrac{8}{5}, \dfrac{31}{5}\right)$

D. $\left(\dfrac{8}{3}, \dfrac{25}{3}\right)$

Answer: C

Solution 1: We need to find a point (x,y) that solves both equations:

$y = 2x + 3$
$y = -0.5x + 7$

So, $2x + 3 = -0.5x + 7$. Then $2x + 0.5x = 7 - 3$. Hence, $2.5x = 4$, which implies that $x = \dfrac{4}{2.5} = \dfrac{4}{\frac{5}{2}} = 4\left(\dfrac{2}{5}\right) = \dfrac{8}{5}$. Then $y = 2\left(\dfrac{8}{5}\right) + 3 = \dfrac{16}{5} + \dfrac{15}{5} = \dfrac{31}{5}$.

Solution 2: Alternatively, we may check each point for both equations. If a point does not satisfy one equation, then it cannot be on the respective line, and there is no need to check the other equation. For completion, we put all the results:

x	$y = 2x + 3$	Correct?	$y = -0.5x + 7$	Correct?
$\dfrac{31}{5}$	$2\left(\dfrac{31}{5}\right) + 3 = \dfrac{62}{5} + \dfrac{15}{5} = \dfrac{77}{5}$	No.	$-0.5\left(\dfrac{31}{5}\right) + 7 = -\dfrac{31}{10} + \dfrac{70}{10} = \dfrac{39}{10}$	No.
$\dfrac{3}{2}$	$2\left(\dfrac{3}{2}\right) + 3 = 3 + 3 = 6$	Yes.	$-0.5\left(\dfrac{3}{2}\right) + 7 = -\dfrac{3}{4} + \dfrac{28}{4} = \dfrac{31}{4}$	No.
$\dfrac{8}{5}$	$2\left(\dfrac{8}{5}\right) + 3 = \dfrac{16}{5} + \dfrac{15}{5} = \dfrac{31}{5}$	Yes	$-0.5\left(\dfrac{8}{5}\right) + 7 = -\dfrac{4}{5} + \dfrac{35}{5} = \dfrac{31}{5}$	Yes.
$\dfrac{8}{3}$	$2\left(\dfrac{8}{3}\right) + 3 = \dfrac{16}{3} + \dfrac{9}{3} = \dfrac{25}{5}$	Yes.	$-0.5\left(\dfrac{8}{3}\right) + 7 = -\dfrac{4}{3} + \dfrac{21}{3} = \dfrac{17}{3}$	No.

54. (Competition 2012) Suppose there are two lines $y = 6x$ and $y = -\dfrac{1}{6}x$, and another line $y = cx$ with $c > 0$ bisecting the angle between these two lines. Find the value of c.

Answer: $\dfrac{5}{7}$

Solution: Let A be the point (1,6), and B be (6,-1). Then A is on the first line, and B is on the second one. Also, if O is the origin (0,0), $\bar{O}A = \bar{O}B = \sqrt{1^2 + 6^2} = \sqrt{37}$. The triangle $\triangle AOB$ is then isosceles, and the angle $A\hat{O}B$ is the one we want to bisect. The line bisecting the angle will also bisect the side AB. The midpoint of AB is $\left(\dfrac{1+6}{2}, \dfrac{6+(-1)}{2}\right) = \left(\dfrac{7}{2}, \dfrac{5}{2}\right)$. Hence we need to find the equation of the line that passes through $(0,0)$ and $\left(\dfrac{7}{2}, \dfrac{5}{2}\right)$. It is $y = \dfrac{5}{7}x$. So, $c = \dfrac{5}{7}$.

55. (Competition 2011) Use the graph below to answer the question that follows.

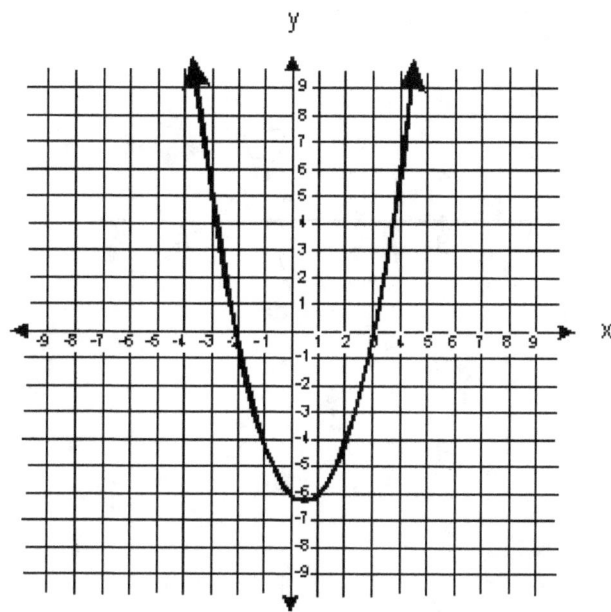

Which equation is represented by this graph?

A. $y = -x^2 + x + 6$

B. $y = x^2 - x - 6$

C. $y = -x^2 - x - 6$

D. $y = x^2 - x + 6$

Answer: B

Solution 1: Parabola passes through the points $(-2, 0)$ (that means $x = -2, y = 0$), $(0, -6)$ and $(3, 0)$. We use that parabolas can be written as $y = ax^2 + bx + c$. For "concave-up" parabolas, we must have $a > 0$, so the answer is either B or D. Plugging $x = 0$, we must have $y = -6$, so we the answer must be B.

Solution 2: Alternatively, we will use that $x = -2$ and $x = 3$ are the roots of the parabola $y = ax^2 + bx + c$. We know that the sum of the roots of a parabola is $Sum = -\frac{b}{a}$, while the product is $Product = \frac{c}{a}$. Since, from the options we have that $a = 1$ (a must be positive), then $b = -1$, and $c = -6$.

56. (Practice 2011) Which of the following equations represents the graph below?

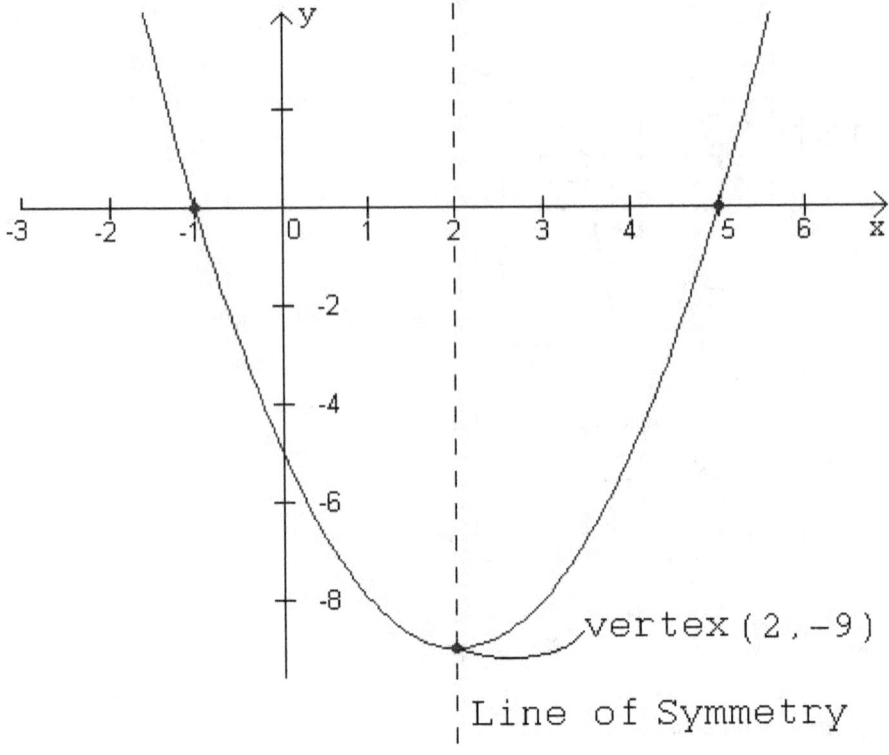

A. $y = x^2 - 5$
B. $y = x^2 - 4x - 5$
C. $y = -x^2 + 5x - 9$
D. $y = x^2 + 4x - 5$

Answer: B

Solution 1: Parabola passes through the points $(-1, 0)$, $(2, -9)$ and $(5, 0)$. We use that parabolas can be written as $y = ax^2 + bx + c$. For "concave-up" parabolas, we must have $a > 0$, so the answer is either A, B or D. Plugging, for instance $x = 5$, we must have $y = 0$, so we the answer must be B.

Solution 2: Alternatively, we will use that $x = -1$ and $x = 5$ are the roots of the parabola $y = ax^2 + bx + c$. We know that the sum of the roots of a parabola is $Sum = -\frac{b}{a}$, while the product is $Product = \frac{c}{a}$. Since, from the options we have that $a = 1$ (a must be positive), then $b = -4$, and $c = -5$.

Chapter 3: Geometry

Measures of angles, of lengths and areas

A circle has 360 degrees or 2π radians. The sum of internal angles of a triangle is $180°$ or π radians.

On a right triangle, Pythagorean Theorem states that the square of the hypotenuse (a) equals the sum of the squares of the other two sides (b and c):

$a^2 = b^2 + c^2$

The circumference of a circle is given by the formula $C = 2\pi r$, where r is the radius.

This is a summary of areas:

Figure	Formula
Circle	$A = \pi r^2$
Triangle	$A = \dfrac{b \times h}{2}$
Rectangle	$A = b \times h$
Square	$A = l^2$

57. (Practice 2013) Consider a triangle ABC, with angles $A\hat{B}C - B\hat{A}C = 50°$. The bisector of angle $A\hat{C}B$ intersects the side AB on the point D. Let E be the point on AC such that $C\hat{D}E = 90°$. Then, what is the measure of the angle $A\hat{D}E$?

Answer: $25°$

Solution:

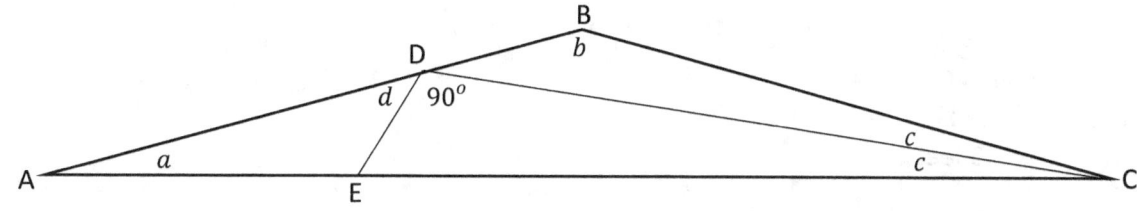

We can write several equations: $b - a = 50°$ (given), $a + b + 2c = 180°$ (sum of internal angles of a triangle), and $d + 90° = b + c$ (external angle of the triangle BCD). Writing b as a function of a on the first equation and substituting on the second and third: $a + (50° + a) + 2c = 180°$ and $d = (50° + a) + c - 90°$. Rewriting, we have: $2a + 2c = 130°$, so $a + c = 65°$. And, $d = (a + c) - 40°$. Finally, $d = (65°) - 40° = 25°$.

58. (Competition 2015) On the triangle ABC, the angle $B\hat{A}C$ is $140°$. Let D be the midpoint of the side BC, E be the midpoint of the side AB, and F be the point on the side AC such that the segment DF is perpendicular to the side AC. What is the angle $E\hat{D}F$, in degrees?

Answer: $90°$

Solution: Let's call θ the angle we need to find (see figure). Observing the quadrilateral $AEDF$, we can write the angle $A\hat{E}D$ as $130° - \theta$. So, the angle $B\hat{E}D$ is $50° + \theta$. Now we use the fact that if we connect the midpoints of two sides of a triangle, then the line is parallel to the third side. So, $50° + \theta = 140°$. Then, $\theta = 90°$

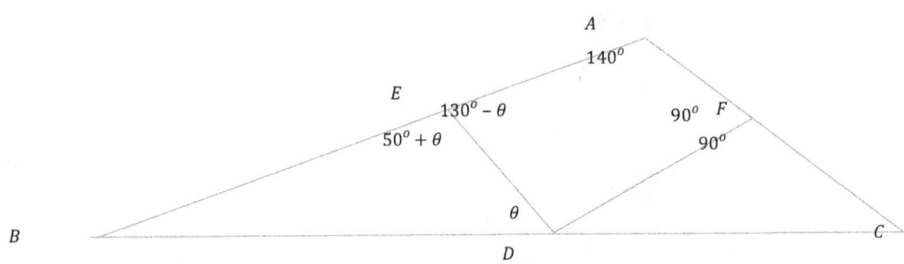

59. (Competition 2011) Use the diagram below to answer the question that follows.

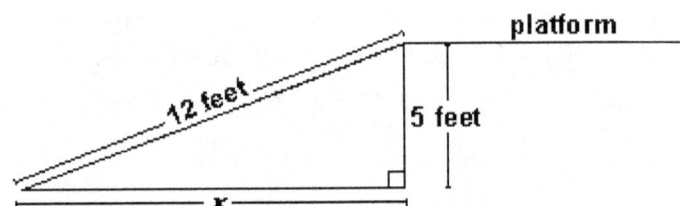

A ramp 12 feet long is leaning against a raised platform which is 5 feet above the ground. What is the distance from the ramp's contact point with the ground and the base of the platform?

A. 7 feet
B. 8.5 feet
C. $\sqrt{119}$ feet
D. 13 feet

Answer: C

Solution: That's an application of Pythagorean Theorem, $a^2 = b^2 + c^2$. So, $12^2 = x^2 + 5^2$. Hence, $144 = x^2 + 25$. Then $x^2 = 119$. The answer is $x = \sqrt{119}\ ft$.

60. (Practice 2012) Two men, starting at the same point, walk in opposite directions for 4 meters, turn left and walk another 3 meters. What is the distance between them?

Answer: 10

Solution: Each man will walk on the legs of a right triangle. At the end, the distance from the initial point will be the hypotenuse of the triangle: $a^2 = 4^2 + 3^2$, so $a = 5$. Since their final positions are collinear with the initial position (draw a sketch of the situation to be convinced), they will be 10 meters apart.

61. (Competition 2012) Given the points A = (-7, 1), B = (2, 6), C = (7, 1) and D = (-2, -4). What term BEST describes the polygon ABCD?

   A. Square;
   B. Rectangle;
   C. Rhombus;
   D. Kite;
   E. Parallelogram
   F. None of previous

Answer: E

Solution: The slope of the line AB is $\frac{6-1}{2-(-7)} = \frac{5}{9}$. Slope of BC is $\frac{1-6}{7-2} = -\frac{5}{5} = -1$. Slope of CD is $\frac{-4-1}{-2-7} = \frac{-5}{-9} = \frac{5}{9}$. Slope of DA is $\frac{1-(-4)}{-7-(-2)} = \frac{5}{-5} = -1$. So, ABCD is a parallelogram. Also, since the intersecting lines are not perpendicular (otherwise the product of their slopes would be -1), then we know that ABCD is not a square or a rectangle. Because a "rhombus" is a parallelogram with all sides equal, we must verify if the adjacent sides have same length or not:
$\bar{AB} = \sqrt{5^2 + 9^2}$
$\bar{BC} = \sqrt{(-5)^2 + 5^2}$ Therefore, ABCD is just a parallelogram.

62. (Competition 2012) How many non-congruent triangles with perimeter 7 will have all its sides with integer length?

Answer: 2

Solution: Let a, b and c be the length of each side of the triangle, with $a \leq b \leq c \leq d$. Then $a + b + c = 7$. Possibilities are: $a = 1, b = 1, c = 5$; $a = 1, b = 2, c = 4$; $a = 1, b = 3, c = 3$; $a = 2, b = 2, c = 3$. However, we need to consider that the sum of two sides of a triangle needs to be greater than the other side. From the possibilities, only 2 satisfy this criterion. Hence, 2 possibilities.

63. (Competition 2012) Let XOY be a right triangle with $X\hat{O}Y = 90°$. Let M and N be the midpoints of OX and OY, respectively. If $XN = 19$ and $YM = 22$, determine the length of the segment $\overline{MN}$.

Answer: 13

Solution: Let $ON = NY = x$ and $OM = MX = y$. We are looking for the value of $\sqrt{x^2 + y^2}$. We know: $x^2 + (2y)^2 = 19^2$ and $(2x)^2 + y^2 = 22^2$. Then: $x^2 + 4y^2 = 361$ and $4x^2 + y^2 = 484$. Adding: $5x^2 + 5y^2 = 845$ so $x^2 + y^2 = 169$. Therefore: $\sqrt{x^2 + y^2} = 13$.

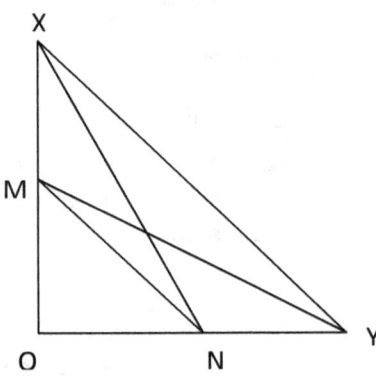

64. (Competition 2012) Suppose we have a regular pentagon with length of each side equal to 1. Find the length of $\overline{BP}$.

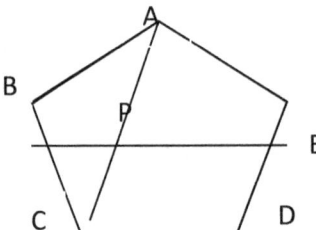

Answer: $\dfrac{-1 + \sqrt{5}}{2}$

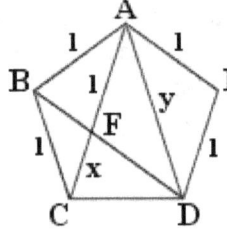

Solution: Each internal angle of the pentagon is 108 degrees. The diagonals split each angle of a regular pentagon into three identical angles. In this diagram, we see that quadrilateral AEDF is a rhombus (parallel sides and consecutive sides are congruent (equal)). And so, the diagonal AD bisects the angle EAF. And, by symmetry, all three angles are congruent (36 degrees).

Use the same diagram, each side of our pentagon is 1. AF=1, because it is the side of our rhombus. Triangles ADF and BCF are similar. So, $\dfrac{x}{1} = \dfrac{1}{y}$. But $y = x + 1$. So $x = \dfrac{1}{x+1}$. Which gives $x(x + 1) = 1$. So $x^2 + x - 1 = 0$. Solving for x, $x = \dfrac{-1 \pm \sqrt{5}}{2}$. Only one of these lengths is positive. So: $x = \dfrac{-1 + \sqrt{5}}{2}$.

65. (Competition 2015) Find the area of the triangle. (Note the area of each small square is 1)

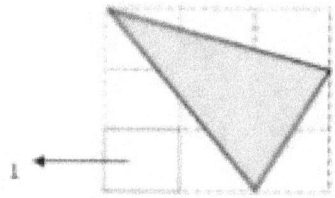

Answer: 3.5 or $\frac{7}{2}$

Solution: The area of the 3×3 square is 9. The area of the white triangle to the left of the shaded one is $\frac{(2)(3)}{2} = 3$. The area of the white triangle on the right bottom is $\frac{(1)(2)}{2} = 1$. The area of the white triangle on the top right is $\frac{(3)(1)}{2} = 1.5$. Hence, the area of the shaded triangle is $9 - 3 - 1 - 1.5 = 3.5$.

66. (Competition 2013 and Competition 2014) Suppose we have the following rectangle by putting together 7 rectangles with perimeters of 42 feet. Find the area of the big rectangle.

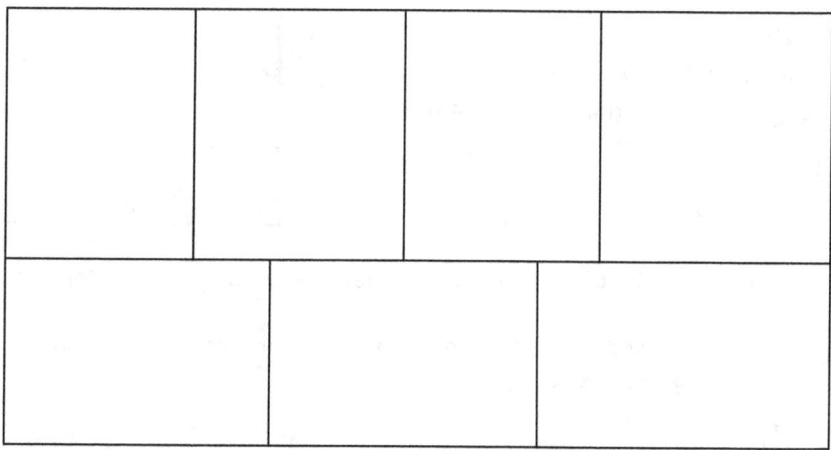

Answer: 756

Solution: Call x the length of the smallest side, and y the length of the other side. Since the perimeter is 42 on each rectangle: $x + y + x + y = 42$, or $x + y = 21$. Also, according to the picture, $4x = 3y$. Then, $\frac{3y}{4} + y = 21$. Hence, $\frac{7y}{4} = 21$. Then $y = 12$, and $x = 9$. Each rectangle has area $(9)(12) = 108$, so the big rectangle is 7 times this area: 756.

67. (Competition 2015) An arrow is formed in a 2 by 2 square by joining the bottom corners to the midpoint of the top edge and the center of the square, according to the picture below.

35

Find the area of the arrow.

Answer: 1

Solution: The total area of the square is $2 \times 2 = 4$. The area of the white triangle below the shaded figure is $\frac{(2)(1)}{2} = 1$. The area of the triangle to the left of the shaded figure is $\frac{(1)(2)}{2} = 1$. The area of the triangle to the right is also 1. Hence the shaded figure has area $4 - 1 - 1 - 1 = 1$.

68. (Competition 2012) A Texan farmer bought a rectangular shaped piece of land with measures 120m by 80m. Due to some environmental laws, he needs to plant trees on 20% of the land. He decides to do so by planting on two strips of equal width, according to the figure below. What is the width of each strip?

Answer: 8

Solution: Total area is
$80m \times 120m = 9600m^2$
Hence 20% of the area is
$0.20 \cdot 9600m^2 = 1920m^2$.
So, each strip is $1920m^2 \div 2 = 960m^2$.
Each width is $960m^2 \div 120m = 8m$.

69. (Competition 2011) Use the diagram below to answer the question that follows.

A window is rectangular with a triangular top section. What is the total area of glass needed for the window?

A. 24 square feet
B. 30 square feet
C. 36 square feet
D. 48 square feet

Answer: B

Solution 1: Let's divide the figure into a triangle and a rectangle. For the triangle: $A_\triangle = \frac{bh}{2} = \frac{(4)(3)}{2} = 6 ft^2$. For the rectangle $A_\square = BH = (4)(6) = 24 ft^2$. So, total area is $30 ft^2$.

Solution 2: Alternatively, create two symmetric trapezoids by connecting the top vertex to the bottom side perpendicularly. For each trapezoid, $A = \frac{(B+b)h}{2} = \frac{(9+6)2}{2} = 15 ft^2$. So, total area is $30 ft^2$.

70. (Competition 2014) The figure on the right is composed of eight circles, seven small circles and one large circle containing them all. Neighboring circles only share one point, and two regions between the smaller circles have been shaded. Each small circle has a radius of 5 cm.

    Calculate the area of the shaded part of the figure.

    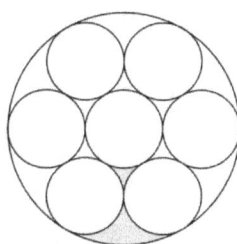

    Answer: $\frac{25}{3}\pi$.

    Solution: If each small circle has radius equal to 5, then the big circle has radius equal to 15. The shaded region is one sixth of the difference of the area of the big circle, $\pi(15)^2$, and the seven small circles, each with area $\pi(5)^2$. So, the answer is $\frac{1}{6}(225\pi - 7 \times 25\pi) = \frac{1}{6}(50\pi) = \frac{25}{3}\pi$.

71. (Competition 2016) Consider the following picture. The circle had radius $r = 3$. What is the area of the square ABCD?

    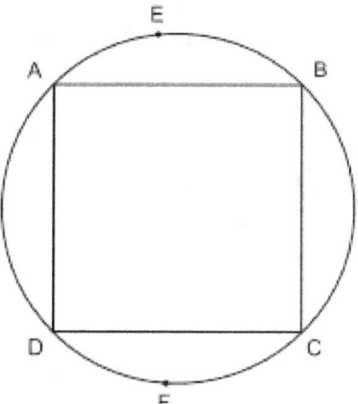

    Answer: 18

    Solution: The diagonal of the square is twice the radius, hence the diagonal is 6. The diagonal is the hypotenuse of an isosceles right triangle, having the side l as both legs: $l^2 + l^2 = 6^2 = 36$. Hence, $2l^2 = 36$, then $l^2 = 18$. And this is exactly the area.

72. (Competition 2016) In the drawing below, the gray rectangle has all its vertices on the equilateral triangle. The area of the triangle is $40\ cm^2$. The smallest side of the rectangle is one quarter of a side of the triangle. What is the area of the rectangle in cm^2?

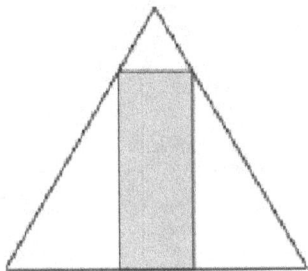

Answer: 15

Solution: If we call l the side of the triangle, then the height will be $\dfrac{l\sqrt{3}}{2}$ (if you do not believe, I recommend drawing an equilateral triangle, drawing a perpendicular line to a side passing through the opposite vertex, this lines cuts the side in half, then use Pythagorean Theorem). So, the area will be $A = \dfrac{l^2\sqrt{3}}{4}$ (area of a triangle is base times height over 2).

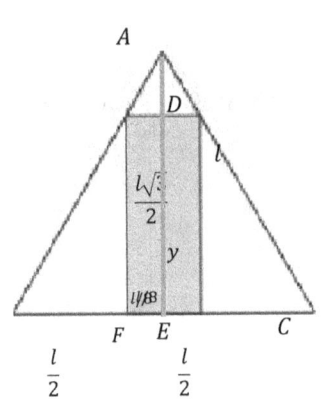

Since the area of the triangle is 40, we can calculate the square of the length of the side: $\dfrac{l^2\sqrt{3}}{4} = 40$, so $l^2 = \dfrac{160}{\sqrt{3}}$.

Now, notice that the triangle AFC is similar to the triangle DEC.

So, $\dfrac{y}{\frac{l\sqrt{3}}{2}} = \dfrac{\left(\frac{l}{2}-\frac{l}{8}\right)}{\frac{l}{2}}$. Then, $y = \dfrac{3\sqrt{3}l}{8}$.

Finally, the area of the rectangle is $\dfrac{l}{4} \times \dfrac{3\sqrt{3}l}{8} = \dfrac{3\sqrt{3}l^2}{32} = \dfrac{3\sqrt{3}}{32} \times \dfrac{160}{\sqrt{3}} = 15$.

73. (Competition 2016) If the graphs of $2y + x + 3 = 0$ and $3y + ax + 2 = 0$ are to meet at right angles, the value of a is:

A. $\pm\dfrac{2}{3}$ B. $-\dfrac{2}{3}$ C. $-\dfrac{3}{2}$ D. 6 E. -6 F. None of previous

Answer: E

Solution: The first line can be rewritten as $y = -\frac{1}{2}x - \frac{3}{2}$, and the second is $y = -\frac{a}{3}x - \frac{2}{3}$. Perpendicular lines will have their slopes in such way that their product is equal to -1:

$$\left(-\frac{1}{2}\right)\left(-\frac{a}{3}\right) = -1$$

Hence, $a = -6$.

74. (Competition 2016) Five points are taken inside or on a square of side 1. Let a be the smallest possible number with the property that it is always possible to select one pair of points from these five such that the distance between them is equal to or less than a. Then a is:

A. $\sqrt{3}/3$ B. $\sqrt{2}/2$ C. $2\sqrt{2}/3$ D. 1   E. $\sqrt{2}$ F. None of previous

Answer: B

Solution: Consider the square of side 1 and divide it into 4 smaller squares of sides $\frac{1}{2}$. By the Pigeonhole principle, one of these smaller squares will contain at least 2 points. And the largest distance within one of the smaller squares is the length of its diagonal: $d^2 = \left(\frac{1}{2}\right)^2 + \left(\frac{1}{2}\right)^2 = \frac{1}{4} + \frac{1}{4} = \frac{1}{2}$.

So, $d = \frac{1}{\sqrt{2}} = \frac{\sqrt{2}}{2}$.

75. (Competition 2014) A circular Ferris wheel has a radius of 8 meters and rotates at a rate of 12 degrees per second. At $t = 0$, a seat is at its lowest point, which is 2 meters above the ground. Determine, in meters, how high above the ground the seat is at $t = 40$ seconds.

Answer: 14

Solution: After 40 seconds, the Ferris wheel traveled $40 \times 12° = 480° = 360° + 120°$. The position of the seat will be at the same height as the $30°$ position.

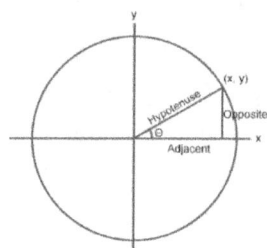

According to the picture on the side, the hypotenuse is $h = 8$ meters. The angle θ is $\theta = 30°$. Using that sine of 30 degrees is $\sin 30° = \frac{1}{2}$. We have that $\sin 30° = \frac{Opposite}{Hypotenuse}$. Hence, the "opposite side" is $Opposite = \frac{1}{2}(8) = 4$ meters.

The height of the seat is the sum of this "opposite" plus the radius plus the 2 meters (minimum height). Answer is $4 + 8 + 2 = 14$ meters.

76. (Competition 2014) The floor of a square hall is tiled with square tiles. Along one diagonal, there are 20 tiles altogether. How many tiles are there on the floor?

Answer: 400

Solution: If there are 20 tiles on the diagonal, then there are 20 tiles on each side. That makes a total of 400 tiles altogether.

Chapter 4: Number Theory

Divisibility Rules and Modular Addition

When is a number divisible by 2? By 3? And so on. There are several rules to determine whether a number can be divided by other. We summarize some:

- 2: if the number ends in an even number;
- 3: if the sum of the digits is a multiple of 3;
- 4: if the last two digits form a number divisible by 4;
- 5: if the number ends in 0 or 5;
- 6: if the number is divisible by both 2 and 3 (see rules above);
- 7: the divisibility rule is not appealing, we recommend dividing and checking;
- 8: if the last three digits form a number divisible by 8;
- 9: if the sum of the digits is a multiple of 9;
- 10: if it ends in 0;
- 11: Add and subtract digits in an alternating pattern (add digit, subtract next digit, add next digit, etc.). Then the answer must be divisible by 11;
- 12: if the number is divisible by both 3 and 4 (see rules above);
- 13: again the divisibility rule is not appealing, we recommend dividing and checking;
- 14: if the number is divisible by both 2 and 7 (see rules above);
- 15: if the number is divisible by both 3 and 5 (see rules above);
- 16: if the last 4 digits form a number divisible by 16;
- 17: again the divisibility rule is not appealing, we recommend dividing and checking;
- 18: if the number is divisible by 2 and 9 (see rules above);
- 19: the divisibility rule is not appealing, we recommend dividing and checking;
- 20: if the last two digits are 00, 20, 40, 60 or 80;
- 21: if the number is divisible by both 3 and 7 (see rules above);
- 22: if the number is divisible by both 2 and 11 (see rules above);
- 23: again the divisibility rule is not appealing, we recommend dividing and checking;
- 24: if the number is divisible by both 3 and 8 (see rules above);
- 25: if the number ends in 00, 25, 50 or 75.

In general, if a number A is divisible by x and by y, and there is no number except 1 that divides both x and y, then this number A will be also divisible by the product xy.

For modular arithmetic, we need to first think of a clock. If the time hand now is at 9, where will it be in 8 hours? Symbolically: $9 + 8 \equiv 5 \ (mod \ 12)$. If the result of the sum is more than 12, we simply take 12, or 24, or 36, or an appropriate multiple of 12 to make the number between 1 and 12.

77. (Competition 2012) Let N be the smallest positive integer such that when multiplied by 33 the result is a number where all digits are 7. What number is N?
(You may leave your answer as a fraction, without reducing it)

Answer: 777777/33

Solution: Since the resulting number is divisible by 3 and 11, we will use the divisibility rules for 3 and 11. The resulting number will only contain 7's, as 3 and 7 are relatively prime, then the quantity of 7's must be a multiple of 3 (in order to the sum of its digits be a multiple of 3). Also, the divisibility rule for 11 is that we have to add the odd-position digits and subtract the even-position digits, the result must be a multiple of 11. If the quantity of 7's is odd, then this rule would give "7", so the quantity of 7's must be even (and this rule will give "0", which is a multiple of 11). The smallest number then would be 777777/33.

78. (Competition 2012 and Competition 2013 with different numbers) Mad at her boyfriend, Safira tore a love letter into n pieces. After that, she took one of the pieces and tore again into n smaller pieces. Since she was still mad, she got one of the last pieces and tore it into n even smaller pieces. From the numbers below, which could represent the final quantity of total pieces? 12, 23, 25, 30 or 38? (Hint: only one value is possible)

Answer: 25

Solution: After the first "breakdown", Safira made n pieces. After the second she had $2n - 1$ (since one piece became n). Finally, she ended up with $3n - 2$ pieces. Hence, we need to find a number such that whenever we add 2, it becomes divisible by 3. The only number satisfying this property is 25.

79. (Competition 2014) How many prime numbers of the form $3^n - 1$, where n is a positive integer, are there between 1 and 2014? (prime numbers are numbers that have only two positive divisors: 1 and themselves)

Answer: 1

Solution: This questions looks much harder than in fact it is. Notice that 3^n is always odd, so $3^n - 1$ is always even. The only even number that is prime is 2, which can be written as $3^1 - 1$. So, only one.

80. (Competition 2013) What is the smallest positive integer greater than 1 that divides the number: $7^{2013} - 13^{20}$?

Answer: 2

Solution: Both numbers are odd, hence their different is even. So the smallest divisor of that number is 2.

81. (Competition 2013) In the sequence BRAZILBRAZILBRAZILBRAZILBRAZIL... Which is the 2013th letter?

Answer: A

Solution: There six letters in BRAZIL. Hence B will be the 1st, 7th, 13th, 19th, ... letters. Similarly, L will the multiple of six: 6th, 12th, 18th, 24th, ... Since $2013 = 6 \times 335 + 3$, then the $(6 \times 335 =)2010$th letter is L. So, 2013th letter will be the third: A.

82. (Competition 2014) In the sequence
MATHEMATICSMATHEMATICSMATHEMATICSMATHEMATICS... Which is the 2014th letter?

Answer: M

Solution: There are 11 letters on MATHEMATICS, letter S will always be on positions multiple of 11: 11, 22, 33, ... Dividing 2014 by 11 we have that: $2014 = 11 \times 183 + 1$. So, last S will be on position $11 \times 183 = 2013$. The next letter, the 2014th, is M.

83. (Practice 2013) In Riverside Pond, 39 ducks are swimming in a straight line. Alex notices that every third duck is white, and the rest are yellow. How many white ducks are there?

Answer: 13

Solution: White ducks will be, 3rd, 6th, 9th, ..., 39th. These are the multiples of 3: $3 \times 1, 3 \times 2, 3 \times 3, ..., 3 \times 13$. So there are 13 white ducks.

84. (Competition 2014) What is $a + b + c$, if each letter represents a digit? (below, "$a\ 7$" represents a two digit number, having a in the tens and 7 in the units, also "$4\ c\ 3$" is four hundred plus c tens plus 3)

$$\begin{array}{r} a\ 7 \\ b\ \times \\ \hline 4\ c\ 3 \end{array}$$

Answer: 15

Solution: The product of b and 7 results on a number having 3 on the unit position. Hence, b can only be 9, and it "carries" 6. Then the product of 9 and a plus 6 equals "forty-c" or $40 + c$. Thus, a can only be 4 and c needs to be a number at least 4 (since $(9)(4) = 36$), or a could be 5 (since $(5)(9) = 45$) with c being less than 5.

Case 1: $a = 4$

Then, the problem would be $47 \times 9 = 423$.

Case 2: $a = 5$

Then, the problem would be $57 \times 9 = 513$, which is not valid.

So, the only possibility is $a = 4, b = 9, c = 2$. The sum is: $a + b + c = 15$.

85. (Competition 2014) A teacher writes a big number on the board and asks a student to tell a divisor of it. The first student says "The number is divisible by 1", the second student says "The number is divisible by 2", the third student says "The number is divisible by 3", and so on. Suppose there are 10 students in the class, and only two statements were incorrect, and these statements were consecutive. What is the smallest possible number written on the board?

Answer: 120

Solution: The first statement is definitely true; every number is divisible by 1. The second one must be true, otherwise the number would be odd so not divisible by 2, 4, 6, 8, or 10, and there would be five wrong statements. Similarly, the number must be divisible by 3, otherwise it would not be divisible by its multiples 3, 6, or 9. If the number is divisible by 2 and 3, it will be divisible by 6. We are trying to get the least common multiple of 8 of the following numbers: (bold numbers cannot be excluded)
$1, 2, 3, 2 \times 2, 5, 2 \times 3, 7, 2 \times 2 \times 2, 3 \times 3, 2 \times 5$

If we exclude 2×2, we must also exclude $2 \times 2 \times 2$, and our number would be:
$2 \times 3 \times 3 \times 5 \times 7 = 630$.

If we exclude 5, we also exclude 2×5, to get: $2 \times 2 \times 2 \times 3 \times 3 \times 7 = 504$. If we don't exclude 5, then we can't exclude $10 = 2 \times 5$.

If we exclude 7, we could also exclude $2 \times 2 \times 2$ (basically we'd exclude an extra 2), or 3×3 (basically we'd exclude an extra 3, making this option smaller). Excluding 9, we'd get:
$2 \times 2 \times 2 \times 3 \times 5 = 120$.

Finally, the last option would be excluding $2 \times 2 \times 2$ and 3×3. The number would be $2 \times 2 \times 3 \times 5 \times 7 = 420$.

86. (Competition 2015) For how many positive integers n is the number $\dfrac{n}{100-n}$ also a positive integer?

Answer: 8

Solution: We need to analyze. Going from 1 to 99, the numbers are: $\dfrac{1}{99}, \dfrac{2}{98}, \dfrac{3}{97}, \ldots$ The first possibility is whenever $n = 50$. Then, for $n = 75$, we'd have $\dfrac{75}{25} = 3$. The other possibilities are $80, 90, 95, 96, 98, 99$.

87. (Competition 2015) Which of the following number divides the number $3^5 \cdot 4^4 \cdot 5^3$?

A. 42   B. 45   C. 52   D. 85   E. 105

Answer:

Solution: We are trying to find a number whose prime decomposition has no other prime other than $2, 3, 5$. And that the power of 3, if it appears on the decomposition, is at most 5, power of 2 is at most 8 ($4^4 = 2^8$), and the power of 5 is at most 3. For completion we put the decomposition of all the numbers: $42 = 2 \times 3 \times 7$ (we can't have 7), $45 = 3^2 \times 5$ (possible), $52 = 2^2 \times 13$ (we can't have 13), $85 = 5 \times 17$ (we can't have 17), $105 = 3 \times 5 \times 7$ (we can't have 7).

88. (Competition 2015) What are the last digit of the number 2^{2015}?

Answer: 8

Solution: We know that $2^1 = 2$, $2^2 = 4$, $2^3 = 8$, $2^4 = 16$, $2^5 = 32$, $2^6 = 64$, ... So, the pattern of the last digit is $2, 4, 8, 6, 2, 4, 8, 6, 2$... Since $2015 = 503 \times 4 + 3$. So, the last digit will be 8.

89. (Practice 2013) If we multiply all positive integers less than or equal to 2014, not multiples of 5, what will be the unit digit?

Answer: 6

Solution: Divide the product into groups of 4 consecutive integers:
$(1 \times 2 \times 3 \times 4) \times (6 \times 7 \times 8 \times 9) \times ... \times (2011 \times 2012 \times 2013 \times 2014)$. Notice that $1 \times 2 \times 3 \times 4$, ends in 4. Also, $6 \times 7 \times 8 \times 9$ also ends in 4. And the last digit of powers of 4 will alternate between 4 and 6: $4, 16, 64, 256, 1024, ...$ Hence, we need to know how many groups of 4 there: if is odd (it'd end in 4) or if it is even (it'd end in 6). The even-numbered groups all end in numbers ending in 9. So, the quantity of groups is odd. Hence, the last digit of the product is 6.

90. (Competition 2016) A sequence of number is created by having the first term equal to $a_1 = 18$, then having the next terms to be equal to the previous terms plus sum of its digits. For instance, $a_2 = 18 + 1 + 8 = 27$. What is NOT a term in this sequence?

    A. 54;   B. 180;   C. 316;   D. 693;   E. 1224.

Answer: C

Solution: We do not need to compute all the numbers, simply notice that since the first number is a multiple of 9, then the sum all the its digits will also be a multiple of 9, making the second number also a multiple of 9. This way, all numbers will be multiples of 9. The only number that is not a multiple of 9 is 316.

91. (Competition 2015) Professor Teixeira has two watches, both shows digital time in a 24-hour format. Both of them are defective. One goes on twice faster than normal time. Other goes backwards at normal speed. Today, both correctly showed $13:00$ at the same time. What will the correct time be at the moment that both watches show the exact same time again?

Answer: 9pm or 21 hours.

Solution 1: If t is less than $\frac{24-13}{2}=5.5$, after t hours, one watch will represent $13+2t$ and the other will represent $13-t$. If t is more than 5.5 but less than 13, than the faster clock will represent $13+2t-24=2t-11$, and the backwards clock will still represent $13-t$. If it is more than 13 but less than 17.5, the faster clock will represent $2t-11$, and the backwards clock will represent $13-t+24=37-t$. We try to see possible solutions:

Time after 13pm	Faster clock	Backwards clock	Possible solution
$0 \le t < 5.5$	$13 + 2t$	$13 - t$	$13 + 2t = 13 - t$ $t = 0$ Not the solution.
$5.5 \le t < 13$	$13 + 2t - 24 =$ $2t - 11$	$13 - t$	$2t - 11 = 13 - t$ $t = 8$ Yes, this is the solution.

So, it will be 8 hours after 13pm, so 9pm. On the faster clock, it would have passed 16 hours, so it would show 5am, while the backwards clock would show 8 hours before 13pm, so also 5am.

Solution 2: Everything would be much simpler using modular arithmetic. We need:

$$13 + 2t = 13 - t \pmod{24}$$
$$3t = 0 \pmod{24}, \text{ that means that } 3t \text{ is a multiple of } 24)$$
$$3t = 24, \text{ so } t = 8.$$

92. (Competition 2015 and Competition 2016 with slightly different numbers) Park wants to burn a CD. He wants to include six songs. The songs' lengths are 7:55, 9:40, 9:15, 12:45, 8:20, and 11:30. What is the total playing time?

Answer: 59min 25sec

Solution: We could add the minutes and seconds separately. Minutes: there are $7+9+9+12+8+11=56$ minutes. Seconds: $55+40+15+45+20+30=205$ seconds, which represents 3 minutes and 25 seconds. The total playing time is 59 minutes and 25 seconds.

Chapter 5: Counting Techniques

Permutations, Combinations, Probability, Counting Techniques

The Fundamental Principle of Counting, also known as the multiplication rule, states that if one task can be performed in n_1 ways, a second task can be performed in n_2 ways, a third task can be performed in n_3 ways, and so on. Then, the total amount of possibilities that these tasks can be performed is

$$n_1 \times n_2 \times n_3$$

If there are more than three tasks, simply adapt the formula.

We define the factorial of the number by:

$0! = 1$

$1! = 1$

$2! = 2 \times 1$

$n! = n \times (n-1) \times \cdots \times 2 \times 1$

In some tasks, order can be important or not. This gives range to two different possibilities:

- Permutation: if the order or position is important.
- Combination: if the order or position is NOT important.

To a permutation, the group ABC is different from the group BCA. For example: passwords, ways to arrange letters, possibilities of giving gold/silver/bronze medals, etc.

To a combination, the group ABC is the same as the group BCA. For example: creating a committee with three people, choosing three people to give tickets, etc.

Formulas: in most cases of direct permutation or combination, there will be a large group with n total individuals, and you will have to choose p possibilities ($p \leq n$):

Permutation: $P_{n,p} = \dfrac{n!}{(n-p)!}$. However, if there is repetition, we should divide by each repeated quantity factorial (see questions below).

Combinations: $C_{n,p} = \dfrac{n!}{p!(n-p)!}$.

For probability, we only use the basic, that is, the probability that some "event" happens, in an experiment with "equally likely outcomes", is the quantity of favorable events, divided by the total amount of possible outcomes.

Finally, some problems will require a really clever and well-organized way of counting the outcomes.

93. (Competition 2012) How many different ways can you rearrange the letters in *school*?

Answer: 360

Solution: It is a combination/permutation problem. Since the order of the objects matters, it will be a permutation problem. If all the letters were different, the answer would be $6! = 6 \cdot 5 \cdot 4 \cdot 3 \cdot 2 \cdot 1 = 720$ permutations. Also, since there is a repetition (there are two letters "o"), changing "o" with "o" does not produce a new rearrangement. Hence we must divide by this number by $2! = 2 \cdot 1 = 2$. The answer is 360.

94. (Practice 2012) How many different ways can you rearrange the letters in *tests*?

Answer: 30

Solution: It is a permutation problem, notice that there are letters being repeated. There are 5 letters, 2 *t*, 2 *s*, and an *e*. The answer is $\dfrac{5!}{(2!)(2!)} = \dfrac{(5)(4)(3)(2)(1)}{(2)(1)(2)(1)} = 30$.

95. (Competition 2016) At a party, 10 people give gifts to each other. How many gifts were exchanged?

Answer: 90

Solution 1: Each person exchanges 9 gifts, so a total of 90 gifts are exchanged.

Solution 2: Each present has two "positions" to be filled: who is giving and who is receiving. There are 10 people to choose to fill these spots, since order is important ("A giving a present to B" is a different action than "B giving a present to A") we are dealing with permutation:

$$P_{10,2} = \dfrac{10!}{(10-2)!} = \dfrac{10!}{8!} = \dfrac{10 \times 9 \times 8 \times 7 \times 6 \times 5 \times 4 \times 3 \times 2 \times 1}{8 \times 7 \times 6 \times 5 \times 4 \times 3 \times 2 \times 1} = 10 \times 9 = 90.$$

96. (Practice 2012) Kaci has 8 pair of jeans, 4 skirts, 12 blouses, and 5 pairs of shoes. How many different outfit combinations does she have? (she can use either a pair of jeans or a skirt, never both at the same time)

Answer: 720

Solution: It is a "counting problem", we will use the multiplication principle. Kaci has 5 options of shoes, 12 options of jeans/skirts, and 12 options of blouses. Hence the total possible combinations is $(5)(12)(12) = 720$.

97. (Practice 2012) Meghan, Owen, Travis, Natalie, and Ana won a set of 3 tickets to a concert. How many different combinations are there for three of the friends to use the tickets?

Answer: 10

Solution: There are 5 friends, but they need to choose only 3. This is a combination problem, since order is not important: $C_{5,3} = \binom{5}{3} = \dfrac{5!}{(3!)(5-3)!} = \dfrac{(5)(4)(3)(2)(1)}{(3)(2)(1)(2)(1)} = 10$.

98. (Competition 2013) Count the number of different ways to reach from A to B, if you can travel on the sides of the rectangle, and only from left to right or upwards.

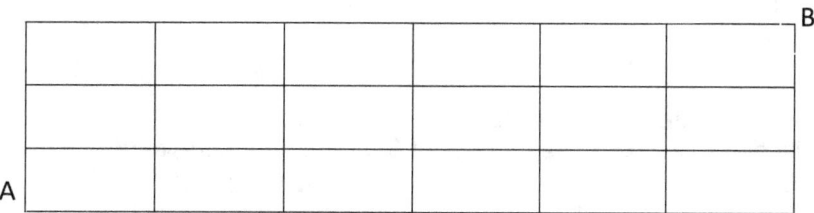

Answer: 10

Solution 1: Let's consider "simpler" versions of the problem

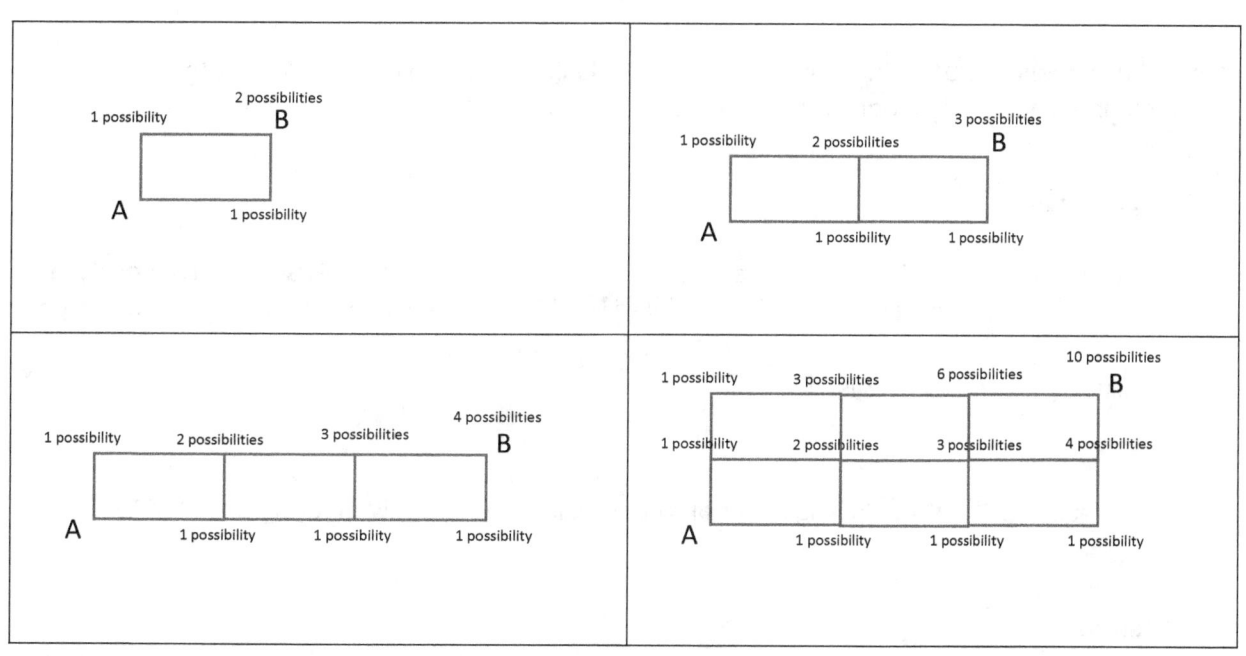

On a general inside rectangle:

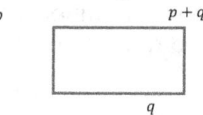

Completing each vertex, we see that there are 84 possibilities.

Solution 2: From A to B, any path would go through 9 edges; we need to choose 3 of these edges to be vertical. Hence, $C_{9,3} = \dfrac{9!}{(3!)(9-3)!} = \dfrac{(9)(8)(7)(6)(5)(4)(3)(2)(1)}{(3)(2)(1)(6)(5)(4)(3)(2)(1)} = 84$.

99. (Practice 2012) Twenty-five index cards numbered 1-25 are placed in a box and drawn at random as students enter the classroom, once a card is drawn, it is discarded. What is the probability that the second student will draw a multiple of 3?

Answer: $\dfrac{8}{25}$.

Solution: There are eight multiples of 3 in the cards: 3, 6, 9, 12, 15, 18, 21, and 24. There are two possibilities: the first card is a multiple of 3, or not. If the first card is a multiple of 3 and the second is also a multiple of 3, the probability of such situation is $\left(\dfrac{8}{25}\right)\left(\dfrac{7}{24}\right) = \dfrac{7}{75}$. On the second possibility (first card is not a multiple of 3, but the second card is): $\left(\dfrac{17}{25}\right)\left(\dfrac{8}{24}\right) = \dfrac{17}{75}$. Hence the answer will be $\dfrac{7}{75} + \dfrac{17}{75} = \dfrac{24}{75} = \dfrac{8}{25}$.

100. (Competition 2014) The numbers 1 to 500 inclusive are put into a hat. What is the probability that a number chosen at random from the hat is a multiple of both 6 and 13?

Answer: $\dfrac{3}{250} = 1.2\%$

Solution: A multiple of 6 and 13, since these number have no common divisor, is also a multiple of $6 \times 13 = 78$. These multiples are: $78, 156, 234, 312, 390, 468$. Hence the probability of random getting a multiple of both 6 and 13 is $\dfrac{6}{500} = \dfrac{3}{250} = 1.2\%$.

101. (Practice 2012) When rolling a pair of dice what is the probability of rolling a total of 7?

Answer: $\dfrac{1}{6}$

Solution: We are dealing with two dice, call them "first die" and "second die". There are 6 possible outcomes on the first die, and 6 possible outcomes on the second die. That makes a total of 36 possible outcomes. There are six different ways of getting 7 (the first number represents the outcome on the "first die", the second number is the number on the "second die"): (1,6), (2,5), (3,4), (4,3), (5,2), and (6,1).

102. (Practice 2012) If 5 coins are tossed simultaneously, what is the probability of getting exactly 3 heads?

Answer: $\dfrac{5}{16}$

Solution: Represent by HHHTT, the event of getting heads (H) on the first 3 attempts, and tails (T) on the last two. The probability of HHHTT is $\left(\dfrac{1}{2}\right)\left(\dfrac{1}{2}\right)\left(\dfrac{1}{2}\right)\left(\dfrac{1}{2}\right)\left(\dfrac{1}{2}\right) = \dfrac{1}{32}$. But this is not the only possibility, we could also have HHTHT, HHTTH, HTHHT, and so on. Each different configuration would also have the exact same probability of $\dfrac{1}{32}$. There are 5 trials, we need to choose 3 to be heads, so there are $C_{5,3} = \binom{5}{3} = \dfrac{5!}{(3!)(5-3)!} = \dfrac{(5)(4)(3)(2)(1)}{(3)(2)(1)(2)(1)} = 10$. Hence the answer is $10\left(\dfrac{1}{32}\right) = \dfrac{5}{16}$.

103. (Practice 2012) If you count from 1 to 100, how many 7's will you pass on the way?

Answer: 20

Solution 1: There are 10 numbers 7 on the unit position: 7, 17, 27, 37, 47, 57, 67, 77, 87, 97. There are 10 numbers 7 on the ten position: 70, 71, 72, 73, 74, 75, 76, 77, 78, 79. Hence 20.

Solution 2: Consider all numbers from 00 to 99, allowing zero to be on the ten position too. There are 100 different numbers, each number with two digits. So there are 200 digits. Each one of the ten digits appears exactly the same amount of time: 20.

104. (Competition 2016) How many pairs of positive integer numbers are there such that their sum is 2016?

Answer: 1007

Solution: Call the smallest number x and the other number y. If we start with $x = 1$ then $y = 2015$, $x = 2$ and $y = 2014$, ... $x = 1007$ and $y = 1009$. But the next pair would not be valid, since both numbers would be 1008. So there are 1007 possible pairs.

105. (Practice 2012) How many four sided figures are in this diagram?

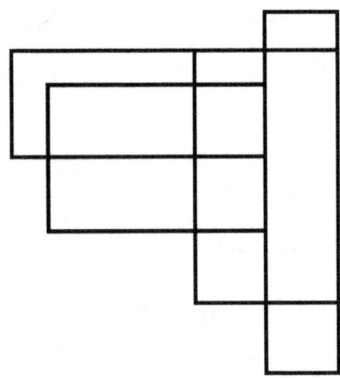

Answer: 25

Solution: One way to solve this problem is finding a good systematic way, Here is one example:

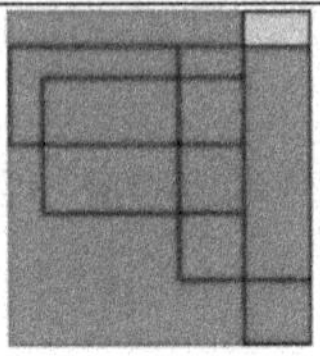 First Picture: 1 rectangle	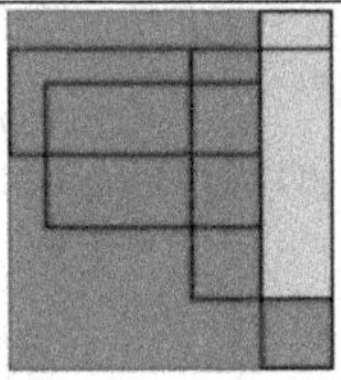 Second Picture: 2 additional rectangles. The small rectangle, which has been added and the big one, which contains the two small rectangles.
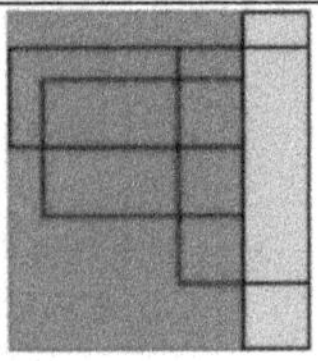 Third picture: 3 additional. The big rectangle. Then two rectangles, which contains 2 small linked rectangles. And the small square, which has been added.	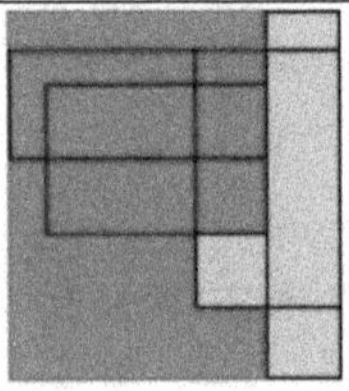 Fourth picture: Only one additional small rectangle.
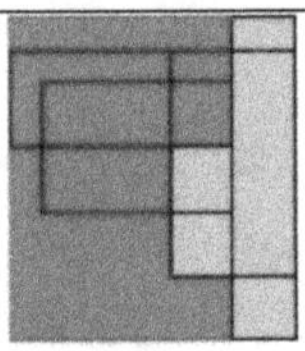 Fifth Picture: 2 additional. The rectangle, which contains the two small rectangle and the small additional rectangle.	Sixth picture: 3 additional. Seventh picture: 5 additional. Eighth picture (adding bottom left): 2 additional Ninth picture: 4 additional Tenth picture: 2 additional

106. (Competition 2014) How many triangles are in this figure?

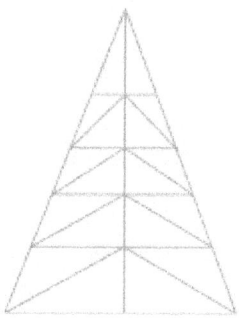

Answer: 31

Solution: Just considering the "top" part, there are 3 triangles. Every time a new "section" is added right below it, 7 new triangles are added. The answer is, then: $3 + 7 + 7 + 7 + 7 = 31$.

107. (Competition 2015) How many triangles are in a fully connected pentagon?

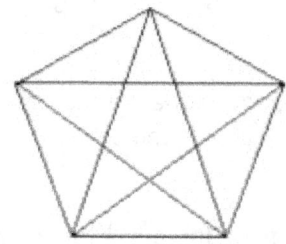

Answer: 35

Solution: There is a "rotational symmetry". So, let's count only one position:

Since there are 5 sets of triangles like these:

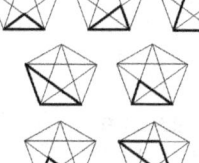
There are $5 \times 7 = 35$ different triangles.

108. (Competition 2016) Draw a picture by phases. On phase 1, draw one hexagon. On phase 2, add congruent hexagons around the existing hexagons according to the picture, making a TOTAL of 7 hexagons. Then, on phase 3, continue the process, and get a TOTAL of 19 hexagons. How many TOTAL hexagons will you have drawn when you complete phase 10.

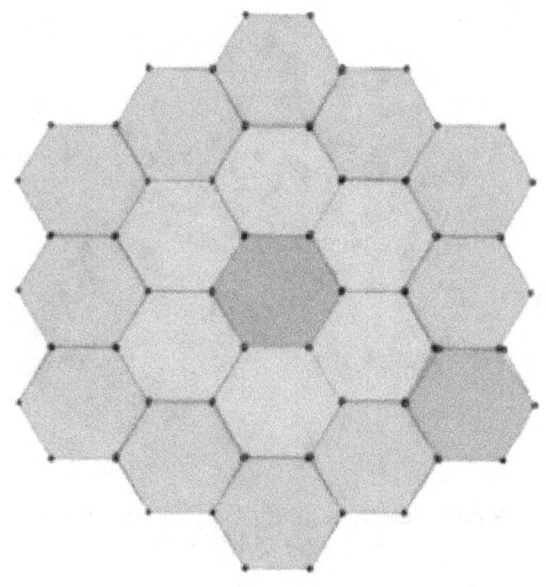

Answer: 271

Solution: Over the red hexagon, we draw lines coinciding with the biggest three diagonals. These lines divide the figure into six congruent parts. See below:

Let's forget the red hexagon for a minute, and count how many hexagons we can count on each step.

Phase 2 (orange triangles), there is 1 hexagon on each part.

Phase 3 (blue triangles), there are $\frac{1}{2} + 1 + \frac{1}{2} = 2$ hexagons on each part.

Phase 4 (not drawn, but easy to imagine), there would be 3 hexagons on each part.

Continuing, we conclude that there will be 9 hexagons on each part when you draw phase 10. Hence the total number of hexagons will be $6(1 + 2 + 3 + ... + 9) + 1 = 6(55) + 1 = 271$.

109. (Competition 2016) The following picture was made with 40 matches. It created a 4 by 4 grid of small squares. Suppose now you have 2016 matches and want to create a bigger n by n grid of small squares. What is the biggest n?

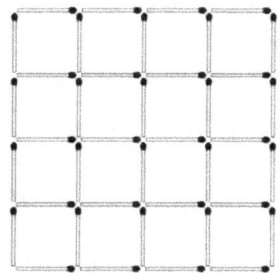

Answer: 31

Solution: In the 4 by 4 grid, we notice that there are 5 rows of 4 horizontal matches, and 5 columns of 4 vertical matches. In an n by n grid, we'd have $(n+1)$ rows of n horizontal matches, and $(n+1)$ columns of n vertical matches. A total of $2n(n+1)$ matches (notice that when $n=4$ this results in 40 matches). We need to solve the inequality $2n(n+1) \leq 2016$. Or $n(n+1) \leq 1008$. We can solve it by trial and error, since we know that $30^2 = 900$, we will start by looking from $n=30$ on. If $n=30$, the expression becomes $30(31) = 930$, so we can go up a little bit. If $n=31$, then $31(32) = 992$, almost there. But if we try $n=32$, then $32(33) = 1056$, too much. So, the best n is 31.

110. (Competition 2016) Ramsey's Theory. Consider a regular polygon of n sides. Draw all the diagonals in it, and then paint each side and each diagonal. You have only two colors: red and blue. What is the minimum number of sides that a polygon must have to guarantee that there is at least one triangle with sides of the same color?

Answer: 6

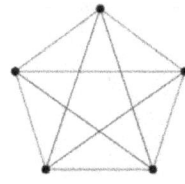

Solution:

We try a pentagon, but we realize that there is no triangle with the same color.

However, if we try an hexagon, we realize that no matter how we paint the sides, there always going to be a monochromatic triangle.

111. (Competition 2016) How many rectangles can you find on the following drawing (hint: squares are also rectangles)?

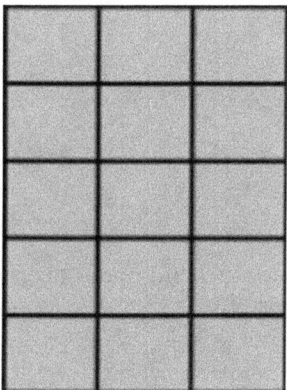

Answer: 60

Solution: We need to find a systematic way of counting the rectangles. Let's start counting how many are 1 by 1, then how many are 1 by 2, and so on. Look at the table:

1 by 1	15
2 by 1	10
3 by 1	5
1 by 2	12
2 by 2	8
3 by 2	4
3 by 3	3
3 by 4	2
3 by 5	1

A total of $15 + 10 + 5 + 12 + 8 + 4 + 3 + 2 + 1 = 60$.

112. (Competition 2015) How many numbers on the list 100, 101, 102, 103, …, 998, 999 do not have digits 2, 5, 7 or 8?

Answer: 180

Solution 1: It can be a counting problem. We are looking for 3-digit numbers not having $2, 5, 7, 8$.

1st digit	2nd digit	3rd digit
Could be $1,3,4,6,9$	Could be $0,1,3,4,6,9$	Could be $0,1,3,4,6,9$
5 possibilities	6 possibilities	6 possibilities

Taking the product: $5 \times 6 \times 6 = 180$.

Solution 2: We can also count.

- From 100 to 109: 6 numbers,
- From 110 to 119: 6 numbers,
- From 120 to 129: no numbers
- From 130 to 139: 6 numbers
- From 140 to 149: 6 numbers
- From 150 to 159: no numbers
- From 160 to 169: 6 numbers
- From 170 to 179: no numbers
- From 180 to 189: no numbers
- From 190 to 199: 6 numbers

Hence, from 100 to 199, there are 36 numbers. But from 200 to 299 there are no numbers (all have 2). Similarly, there are 36 numbers between 300 and 399, 36 more from 400 to 499, 36 more from 600 to 699, and finally, 36 more from 900 to 999. A total of $5 \times 36 = 180$.

113. (Competition 2016) How many possible ways are there to select three positive integers n_1, n_2, n_3 such that the following conditions are met?

(1) Each number is drawn from 1 to 5. That is,
$1 \leq n_1 \leq 5$,

$1 \leq n_2 \leq 5$,

$1 \leq n_3 \leq 5$.

(2) And $n_1 \geq n_2 \geq n_3$.

Answer: 35

Solution: In the first step, this problem can be converted to a "Balls and Urns" problem in the following:

Consider that we have 5 urns, labeled U 1 , U 2 , ... , U 5 . We throw 3 balls into these five urns. The number of combinations in the original problem is the same as the number of ways of distributing 3 balls in 5 urns. For example, if we have two balls in U 5 and one ball in U 3 and all other urns are empty, this corresponds to a solution of n 1 =5, n 2 =5, n 3 =3 .

In the second step, the "Balls and Urns" problem can be converted to a "Stars and Bars" problem. The general question is: How many ways are there to put m identical balls in n urns? To represent each possible partition of m balls into n urns, we can put all the balls in a line, and use n-1 bars as separators. These n-1 separators will divide all the balls into n urns. To do this, we can make m+n-1 stars in a line, and choose n-1 stars out of m+n-1 stars to turn them into bars or separators. The rest of the stars will represent the balls. So there are (m+n-1 n-1) ways of such combinations.

So, back to our original problem. We put 3 balls into 5 urns. That is to select 4 bars out of 7 stars (3 balls plus 4 separators). The total number of ways is (74)=(73)= 7 · 6 · 5 1 · 2 · 3 =35.

Chapter 6: More Math

Patterns

In Mathematics, patterns play a really important role. Number patterns — such as 3, 6, 9, 12 — are very familiar to us. But patterns are much broader. They can be sequential, spatial, temporal, and even linguistic.

The National Council of Teachers of Mathematics recognizes the importance of patterns in its publication Curriculum and Evaluation Standards for School Mathematics:

> *Patterns abound in our world. The mathematics curriculum should help sensitize students to the patterns they meet every day and to the mathematical descriptions or models of these patterns and relationships. (pp. 100/101)*

Problems here were inspired by IQ tests, where the observation of a pattern is tested.

We also included some interesting problems whose classification did not meet any of previous chapters.

114. (Practice 2012) Fill in the missing number: 0,1,1,2,3,5,8,13,__,34,55

Answer: 21

Solution: This sequence is called Fibonacci sequence, each number, from the 3rd on, is the sum of the previous two numbers: $8 + 13 = 21$.

115. (Practice 2012) Following the pattern shown in the number sequence below, what is the missing number?

1 8 27 __ 125 216

Answer: 64

Solution: The sequence is composed by the cubes of natural numbers: $1^3, 2^3, 3^3, 4^3, 5^3, 6^3$.

116. (Competition 2014) Complete the sequence

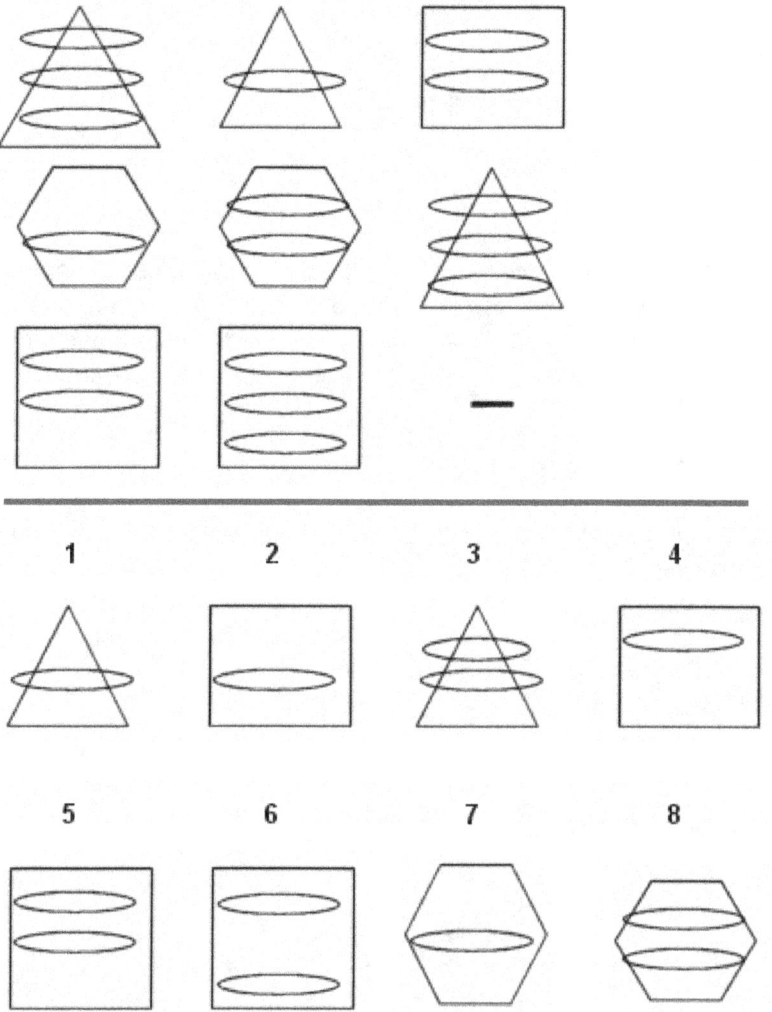

Answer: 7

Solution: It is just a matter of recognizing the pattern: every column has one triangle, one square, and one hexagon. So, it is missing a hexagon. So, we are between 7 or 8. Also, each column must have 1 oval, 2 ovals, and 3 ovals. So, it is missing 1 oval. The answer is 7.

117. (Practice 2012) Which of the figures below the line of drawings best completes the series?

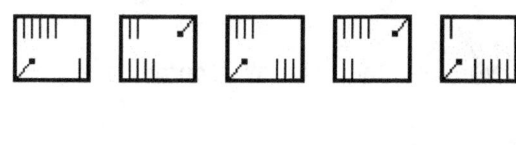

A.

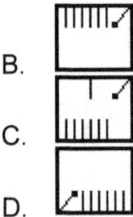
B.

C.

D.

Answer: B

Solution: We need to identify patterns. First, we notice that there is an alternating pattern on the "match-like" piece: bottom-left, top-right, bottom-left, top-right, bottom-left, so on the next one, this piece will be on top-right. So we are between B or C. Finally, we notice that the side that contains the "match-like" piece always gets one extra stick every step: 1, 2, 3, 4, 5, so the next drawing we would see 6 sticks on the side of the match.

118. (Practice 2012) Sally likes 225 but not 224; she likes 900 but not 800; she likes 144 but not 145. Which does she like 1600 or 1700?

Answer: 1600

Solution: Sally likes square numbers, $225 = 15^2$, $900 = 30^2$, $1600 = 40^2$.

119. (Practice 2012) Which vowel comes midway between J and T?

Answer: O

Solution: J K L M N O P Q R S T.

120. (Competition 2016) If a light beam is fired through the top left corner of a 2 by 3 rectangular prism block at an angle of 45 degrees towards the opposite wall, it will emerge from the top right corner (in this experiment, light beam will emerge if it hits one of the corners), see picture on the left.

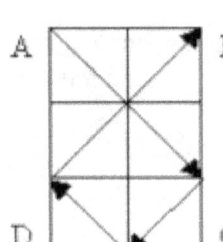
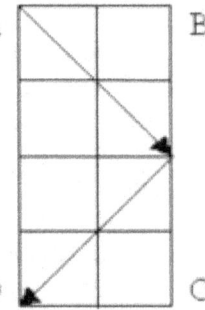

If the light bounces around a 2 by 4 rectangle, it will emerge from the bottom left corner, see picture on the right.

From which corner will the light beam emerge in a 2 by 2016 rectangle?

A. Top left;
B. Top right;
C. Bottom left;
D. Bottom right;
E. It will never emerge.

Answer: C

Solution: Notice that if n is even, then the light beam will emerge from one of the bottom corners. The bottom right will be whenever n is $2, 6, 10, 14, 18, \ldots$ And through bottom left if n is $4, 8, 12, 16, \ldots$ So, we identify that for multiples of 4, the answer will be "bottom left". Since 2016 is a multiple of 4, then the answer is C.

121. (Competition 2013) Let $S_1 = 1$, $S_2 = 1 - 2$, $S_3 = 1 - 2 + 3$, $S_4 = 1 - 2 + 3 - 4$, $S_5 = 1 - 2 + 3 - 4 + 5$, and so on (odd numbers are added, even numbers are subtracted). What is the value of S_{57}?

Answer: 29

Solution: We have $S_1 = 1$, $S_2 = -1$, $S_3 = 2$, $S_4 = -2$, $S_5 = 3$, $S_6 = -3$, $S_7 = 4$. Hence if n is odd, then $S_n = \dfrac{n+1}{2}$. So, $S_{57} = 29$.

122. (Practice 2013) Erin's calculator is broken. Whenever she presses the "square root" button, it randomly executes one of the following two operations: either it calculates the square root (like it was supposed to) or it divides the number by 100. Erin typed the number 201420142014, and pressed the "square root" button several times. How many times she has to press it, in order to guarantee that a number less than 2 is finally shown?

Answer: 8

Solution: Let's try to think about the slowest possibility to decrease. If a number is more than 10,000, then square root will produce a number smaller than the division by 100. So, for numbers greater than 10,000, let's assume that the operation is "division by 100", for numbers less than 10,000, let's assume it takes square root.

Slowest possibility starting with 201420142014: (we know that $16^2 = 256$, and $64^2 = 4096$)

1st operation	2nd operation	3rd operation	4th operation
2014201420.14	20142014.2014	201420.142014	2014.20142014
Division by 100	Division by 100	Division by 100	Division by 100

5th operation	6th operation	7th operation	8th operation
between 16 and 64	between 4 and 8	between 2 and 3	between 1.4 and 2
Square root	Square root	Square root	Square root

123. (Practice 2013) Let's call mirror image of a natural number of TWO digits the number that we get if we change the order of its digits. For instance, the mirror image of 53 is 35. How many two-digit-numbers are there such that the number added to its mirror image is a perfect square?

Answer: 8

Solution: Call the first two-digit number $AB = 10A + B$, with $A \neq 0$. So its mirror will be $BA = 10B + A$, with $B \neq 0$. So, the sum is $10A + B + 10B + A = 11(A + B)$. Since this sum must be a perfect square, we need $A + B = 11$. The possibilities are 29, 38, 47, 56, 65, 74, 83, 92. So, 8 possibilities.

124. (Practice 2013) Let's call mirror image of a natural number of FOUR digits the number that we get if we change the order of its digits. For instance, the mirror image of 5312 is 2135. How many four-digit-numbers are strictly less than its mirror image?

Answer: 4005

Solution: Suppose the number is $ABCD$, with $A \neq 0$. So its mirror is $DCBA$. If $A < D$, then $ABCD < DCBA$. Let's count how many four digit numbers will be strictly less than its mirror dividing into two situations:

Case 1: $A < D$

Then, A cannot be 9, if A is 8, then D has to be 9. B and C could be any number: $8BC9$ there are 100 possibilities.

If $A = 7$, then D could be 8 or 9: $7BC8$ or $7BC9$ would give 200 possibilities. If $A = 6$, then there would be 300 possibilities. If $A = 5$, 400 possibilities. When $A = 1$, more 800 possibilities.

So far, with $A < D$, there would be $100 + 200 + 300 + \ldots + 800 = 3600$.

Case 2: $A = D$

Then our number would look like $ABCA$, and we must have $B \neq C$, so $B < C$. If $B = 8$, then C must be 9. Hence, since A can be any number between 1 and 9, there would be 9 new possibilities.

If $B = 7$, then C could be 8 or 9, so there would be 18 new possibilities. For $B = 6$, we'd have 27 new numbers. If we continue like this, for $B = 0$, there would be 81 numbers.

Total number in this case would be: $9 + 18 + 27 + \ldots + 81 = 405$

The answer for the problem $3600 + 405 = 4005$.

125. (Practice 2013) Consider the function $l(x)$ to be biggest integer less than or equal to x. For example $l(3.4) = 3$ and $l(9) = 9$. What is the value of:
$l(\sqrt{1}) + l(\sqrt{2}) + l(\sqrt{3}) + \cdots + l(\sqrt{50})$?

Answer: 217

Solution:

$l(\sqrt{1})$	$l(\sqrt{2})$	$l(\sqrt{3})$	$l(\sqrt{4})$	$l(\sqrt{5})$	$l(\sqrt{6})$	$l(\sqrt{7})$	$l(\sqrt{8})$	$l(\sqrt{9})$	$l(\sqrt{10})$

1	1	1	2	2	2	2	2	3	3
$l(\sqrt{11})$	$l(\sqrt{12})$	$l(\sqrt{13})$	$l(\sqrt{14})$	$l(\sqrt{15})$	$l(\sqrt{16})$	$l(\sqrt{17})$	$l(\sqrt{18})$	$l(\sqrt{19})$	$l(\sqrt{20})$
3	3	3	3	3	4	4	4	4	4
$l(\sqrt{21})$	$l(\sqrt{22})$	$l(\sqrt{23})$	$l(\sqrt{24})$	$l(\sqrt{25})$	$l(\sqrt{26})$	$l(\sqrt{27})$	$l(\sqrt{28})$	$l(\sqrt{29})$	$l(\sqrt{30})$
4	4	4	4	5	5	5	5	5	5
$l(\sqrt{31})$	$l(\sqrt{32})$	$l(\sqrt{33})$	$l(\sqrt{34})$	$l(\sqrt{35})$	$l(\sqrt{36})$	$l(\sqrt{37})$	$l(\sqrt{38})$	$l(\sqrt{39})$	$l(\sqrt{40})$
5	5	5	5	5	6	6	6	6	6
$l(\sqrt{41})$	$l(\sqrt{42})$	$l(\sqrt{43})$	$l(\sqrt{44})$	$l(\sqrt{45})$	$l(\sqrt{46})$	$l(\sqrt{47})$	$l(\sqrt{48})$	$l(\sqrt{49})$	$l(\sqrt{50})$
6	6	6	6	6	6	6	6	7	7

So, $1 \times 3 + 2 \times 5 + 3 \times 7 + 4 \times 9 + 5 \times 11 + 6 \times 13 + 7 \times 2 = 217$.

126. (Competition 2013) Someone goes from Victoria to Sugar Land with average speed of 70mph and goes back with average speed of 30mph. What was the average speed of the whole trip?

Answer: 42

Solution 1: Let d be the distance from Victoria to Sugar Land, t_1 be the time spent going from Victoria to Sugar Land, and t_2 be time spent on the way back. Average speed is total distance divided by total time ($v = \frac{d}{t}$). We are trying to find $v = \frac{2d}{t_1 + t_2}$. For the first trip: $70 = \frac{d}{t_1}$, which implies $t_1 = \frac{d}{70}$. Similarly, $t_2 = \frac{d}{30}$. Hence the total time was $t_1 + t_2 = \frac{d}{70} + \frac{d}{30} = \frac{3d + 7d}{210} = \frac{10d}{210} = \frac{d}{21}$. Hence the whole trip average speed was $v = \frac{2d}{\frac{d}{21}} = 2d \times \frac{21}{d} = 42 mph$.

Solution 2: A much less elegant way of solving this problem could be by giving a random value to the distance, since the answer will not depend on the actual distance. If we say that $d = 210$ miles (why did we choose 210 miles and not other value?) then it took 3 hours to go, and 7 hours to go back. So a total of 10 hours to complete 420 miles. Hence an average speed of 42mph.

127. (Competition 2013) In a tournament with 2013 teams, on each game there is a loser and a winner, no ties are allowed. Every time a team loses, they are eliminated from the competition. How many games are necessary to determine the champion?

Answer 2012

Solution: Each team that is not the champion lost exactly one game. There is a correspondence between the number of eliminated teams and the number of games. Hence we need 2012 games to determine the champion.

128. (Competition 2014) An enigmatic guy once replied a question about which day of the week was by saying "when the day after tomorrow is yesterday, today will be as far from Sunday as today was from Sunday when the day before yesterday was tomorrow". What day was it?

Answer: Sunday

Solution: When the day after tomorrow is yesterday, today will be three days later, just as when the day before yesterday was tomorrow carries us back three days, which must be Sunday, to be midway between the two "todays".

	Day before yesterday			Day after tomorrow	
Today 2	Tomorrow 2		Real today	Yesterday 1	Today 1

129. (Competition 2015) What is the largest number of tokens we could put on a 5 by 5 chess board, at most one per each cell, such that the number on each row and each column is a multiple of 3?

Answer: 15

Solution: Each column can have at most 3 tokens. So, the maximum number of tokens is $5 \times 3 = 15$. The goal is to produce a configuration having 15 tokens. Here is a possibility (not unique).

130. (Competition 2015) In a game of chess, the Bishop can take any piece that lies on a diagonal.

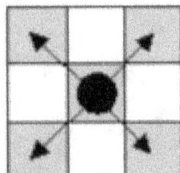

By placing five Bishop pieces on a 4 by 4 grid every square is protected in such a way that no Bishop threatens any other Bishop.

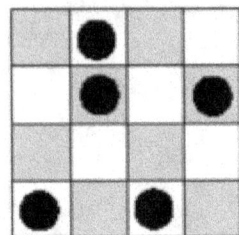

However, it is possible to achieve the same with just four bishops. What is the least number of Bishop pieces required to protect every square on a standard 8 by 8 chessboard?

Answer: 8

Solution: It is possible to protect every square on a chessboard by using just eight Bishop pieces; this is one way it can be done.

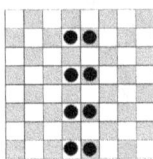

131. (Competition 2015) Four friends, Ali, Brad, Jeff and Park, are playing cards. There are 20 different cards; each one has one color (blue, green, yellow or red). And a number from 1 to 5. Each friend received five cards, so each card belongs to someone. They say:

Ali: "I have four cards with the same number."

Brad: "All my cards are red."

Jeff: "I don't have any blue or green card."

Park: "I have a full house, that is, 3 cards with the same number, and 2 cards with other number."

Only one person is lying. Who lied?

    A. Ali   B. Brad   C. Jeff   D. Park   E. Impossible to find

Answer: B

Solution: Notice that Ali's statement is contradictory with Brad's, since if Brad has five red cards, then Ali's has no red card, meaning he cannot have that alleged number in a red card. Now let's try to find another pair of contradiction. If both Brad and Jeff are telling the truth, then Jeff has all the yellow cards, so Park would not have a full house, since he would not be able to collect 3 cards with the same number (each number, there are only 4 different cards with it, each with a different color). So, Brad is lying.

132. (Competition 2015) Suppose A and B are 1-digit positive integers satisfying:

$(2+1) \cdot (2^2+1) \cdot (2^4+1) \cdot (2^8+1) = 4^A - B$. Find $A + B$.

Answer: 9

Solution 1: This is a very cute solution, we have to just multiply this product by 1. But instead of writing 1, we write $(2-1)$. And we use that $(a-b)(a+b) = a^2 - b^2$.

$(2-1) \cdot (2+1) \cdot (2^2+1) \cdot (2^4+1) \cdot (2^8+1) = (2^2-1) \cdot (2^2+1) \cdot (2^4+1) \cdot (2^8+1) =$
$(2^4-1) \cdot (2^4+1) \cdot (2^8+1) = (2^8-1) \cdot (2^8+1) = 2^{16} - 1 = 4^8 - 1$

So, $A = 8$ and $B = 1$. $A + B = 9$.

Solution 2: We can calculate the product: $(3)(5)(17)(257) = 65{,}535$. But this number is slightly bigger than $(2)(2^2)(2^4)(2^8) = 2^{15}$, the next power of 2 is $2^{16} = 4^8 = 65{,}536$. So, we can write the number as $= 4^8 - 1$. So, $A = 8$, $B = 1$.

133. (Competition 2014) A binary number is a sequence of zeros and ones. How many binary numbers of 8 digits (also called an 8-bit) can you produce if you cannot have two digits 1 next to each other? For example, 00010100 is valid, but 00110000 is not.

Answer: 55

Solution: Let's imagine a constructive approach. Start with only one digit. The possibilities are either 0 or 1. For two digits, the only new number that appears is 10, since adding zeros to the left does not create a new number. For visual effect, let's write in bold the new numbers on each step. So, we have $00, 01, \mathbf{10}$. We will create new numbers by adding a digit on the right of each number. The number 00 does not generate any new number. The number 01 generates only 010 but that's already written, while 10 can derive two new numbers: 100 and 101. So, for 3-digits there are: $000, 001, 010, \mathbf{100}, \mathbf{101}$. Now, considering 4-digits: the only numbers that will generate new numbers are the ones created last step. Number 100 generates 1000 and 1001, and 101 generates 1010. So far, there are: $0000, 0001, 0010, 0100, 0101, \mathbf{1000}, \mathbf{1001}, \mathbf{1010}$. For 5-digits, again only the new numbers from previous step can generate new numbers on this step. If the right-most number is 0, then it generates two new numbers (adding either 0 or 1), but if the right-most digit is 1, then we can generate only one extra number, by adding 0. The possibilities are:

00000, 00001, 00010, 00100, 00101, 01000, 01001, 01010, 10000, 10001, 10010, 10100, 10101. Now it is clear that our sequence follows a Fibonacci rule. The total numbers on a step equals the sum of the total numbers on previous step (same numbers, just adding a zero on the left) plus the numbers on two steps before (adding 10 on the left).

One digit: 2	Two digits: 3	Three digits: 5	Four digits: 8
Five digit: 13	Six digits: 21	Seven digits: 34	Eight digits: 55

134. (Competition 2015) Martians just visited UHV this morning! They have six fingers on each hand, so instead of developing a decimal system, they developed a "duodecimal" system, that is, it is base 12, instead of base 10. They wrote that we need to clap our hands $(123)_{12}$ times (the number $(123)_{12}$ is written in base 12). How many times does this represent?

Answer: 171

Solution: More explanation about different bases can be found on "Mathemagics" chapters.
$(123)_{12} = 3 \times 12^0 + 2 \times 12^1 + 1 \times 12^2 = 3 + 24 + 144 = 171$

135. (Competition 2016) You found an extraterrestrial schoolbook. Instead of using a decimal system, they use a system with only 6 symbols: &, !, @, #, $, %. You realized that:
& = 0, ! = 1, @ = 2, # = 3, $ = 4, % = 5, !& = 6,
!! = 7, !@ = 8, !# = 9, !$ = 10, !% = 11, @& = 12

Which number is !#%?

Answer: 59

Solution: This is simply a conversion of base problem. We notice that these ETs use base 6 number system. The question is then to compute $(135)_6$ into decimal. So,
$(135)_6 = 5 \times 6^0 + 3 \times 6^1 + 1 \times 6^2 = 5 + 18 + 36 = 59$.

Chapter 7: "Mathemagics"

We use a decimal numerical system, that is, there are ten symbols to represent quantities: 0, 1, 2, 3, 4, 5, 6, 7, 8, and 9. With them, we create infinitely-many numbers. Why is our system "decimal"?

The most likely answer is because there are 10 fingers in our hands!

Our numerical system is also a "positional" one. That means that each position within a number represents a different quantity. For example, the number 234 represents two 100's, plus three 10's, and plus four 1's. For integers, the number of digits can give a good description about the quantity. For example, a 3-digit number represents a quantity between 100 and 999.

But that did not used to be the rule. In ancient Rome, for instance, numbers did not have the "positional" characteristics that can be seen today. The size of a representation of an integer did not give much info about the size of the quantity itself. The number XXIII is smaller than the number L, for example.

Octal System

Going back to the positional systems, could we create a system with only 8 symbols? The answer is yes! We would follow the same rules that we use in the decimal system, except that we would only use 8 symbols. For simplicity, the eight symbols could be: 0, 1, 2, 3, 4, 5, 6, and 7. For visual effect, let's use a different font: 0, 1, 2, 3, 4, 5, 6, and 7. Even though, these symbols mean exactly the same (the symbol 5 or 5 would both represent the same quantity: five), their construction will be slightly different due to the lack of more symbols.

In the decimal system, we start to count from zero all the way to nine. Once we complete our first ten, we mark a new position (the 10's) with the digit one, and on the 1's position we start again from 0 to 9. If we complete our second group of ten, we add one to the position of the 10's and start over again. After completing nine 10's, if we complete another group of ten, we need to create a new position (the 100's) and so on...

Back to our "octal" system: for every group of eight, we add one to the position on the left (let's not give any name to this position so far). So we start at 0, 1, 2, 3, 4, 5, 6, 7, and what would happen next? Since we are completing our first group of eight, we start a new position (on the left) with the number one and start back from zero: 10 represents one group of eight.

Following these rules, 36 represents three groups of 8 plus a group of 6, or $6 + 3 \times 8 = 30$. And the biggest two-digit number we could create would be 77. How big is 77 in our decimal system?

Symbolically, this number represents seven complete groups of 8 plus a group of 7, or $7 + 7 \times 8 = 63$.

What would happen to our representation if we add one to the number 77 in our octal system? We would complete the group of eight in the 1's position, so we would have to add one to the other position (still with no name). But that would max-out the symbols, so we would need to create a new position on the left, and the result would be 77+1=100.

How do we interpret the positions on a three-digit number, for example 234? The right-most position (containing the digit 4) represents ones, the position next to its left (containing the digit 3) represents how many groups of 8, and finally the position to the left-most (digit 2) represents eight groups of eight, or a group of $8^2 = 64$. Hence, converting to decimals it becomes: $4 + 3 \times 8 + 2 \times 8^2 = 160$.

Notice that this is very similar to what we are used to in decimal systems. In our decimal system, from right to left, the positions are: 1's, 10's, 100's, 1000's, and so on. For instance, the number we can decompose: $1{,}234 = 4 \times 1 + 3 \times 10 + 2 \times 100 + 1 \times 1000$. To help us see the pattern, let's write it as:

$1{,}234 = 4 \times 10^1 + 3 \times 10^1 + 2 \times 10^2 + 1 \times 10^3$

If an octal number is written, then we could easily convert to decimal system:

$1{,}234 = 4 \times 8^0 + 3 \times 8^1 + 2 \times 8^2 + 1 \times 8^3 = 668$

Other bases

With a similar idea, we could convert from any system to the decimal system. For instance, if the number $(4512)_6$ represents a quantity in base 6, then this number in the decimal system would be:

$(4512)_6 = 2 \times 6^0 + 1 \times 6^1 + 5 \times 6^2 + 4 \times 6^3 = 2 + 6 + 180 + 864 = 1052$

Could we create a system with more than ten symbols? For illustration, how would we handle a "hexadecimal" (16 symbols) system?

The answer is very simple: there are ten standard mathematical symbols for numbers (from 0 to 9), we would need six extra symbols. How about borrowing symbols from the alphabet? Let's put them in order: 0, 1, 2, 3, 4, 5, 6, 7, 8, 9, A, B, C, D, E, and F. With the symbol A meaning what we understand as "ten", B meaning the quantity "eleven", up to the quantity "fifteen" represented by F. This way, an hexadecimal number could be easily converted to decimal: $9AB = 11 \times 16^0 + 10 \times 16^1 + 9 \times 16^2 = 11 + 160 + 2304 = 2475$.

Binary Digits

Finally, we could think of a numerical system containing only two symbols: 0 and 1. This system is known as binary system, and it was invented by the co-inventor of *Calculus*, Gottfried Wilhelm Leibniz (1646-1716), when he published his work in the 1701 paper *Essay d'une nouvelle science des nombres*, in the Paris Academy. Nowadays, this system has applications in digital encoding, since a computer system can have only two states: OFF (zero) or ON (one).

For instance, the biggest number we could get with six-digit binary number would be:

$(111111)_2 = 1 \times 2^0 + 1 \times 2^1 + 1 \times 2^2 + 1 \times 2^3 + 1 \times 2^4 + 1 \times 2^5 = 1 + 2 + 4 + 8 + 16 + 32 = 63$

So, with at most six digits, we could create any number from zero to sixty-three. For example:

$(100101)_2 = 1 \times 2^0 + 0 \times 2^1 + 1 \times 2^2 + 0 \times 2^3 + 0 \times 2^4 + 1 \times 2^5 = 1 + 4 + 32 = 37$

Magic Trick 01: Binary Cards

Material: Make a copy of the Binary card page on the Appendix. Cut them into individual cards for effect purposes.

Preparation: For organization and to make it easier to refer to the cards, let's number them. The first card will be the one that has number 1 on the top left, second card will be the one with number 2 on top left, third will have 4 on the top left, fourth has 8 on the top left, fifth has 16 on top left, and the sixth and last card has 32 on top left. In general, the card number n will have 2^{n-1} on top left.

Performance: Ask someone to think of a positive integer less than or equal to 63, but not reveal it. Show each card and question whether the person sees the number. Make two piles: one with all the cards the person sees the number, and other with the cards that do not contain the number. Immediately after the person answers about the last card, you will say the chosen number out loud.

Trick: It is a very simple trick, every time a person says that the chosen number is on the card, you mentally add the number on the top left of the card. Add while you show the cards, so the magical effect happens faster.

Explanation: Each card represents a position in a 6-digit binary number: $(abcdef)_2$. Remember that each letter can only be one of two possibilities: either 0 or 1. And since this number is written in base 2, it could be easily converted to decimals:

$$(abcdef)_2 = f \times 2^0 + e \times 2^1 + d \times 2^2 + c \times 2^3 + b \times 2^4 + a \times 2^5$$
$$= f \times 1 + e \times 2 + d \times 4 + c \times 8 + b \times 16 + a \times 32$$

Since each letter can be only 0 or 1, you simply add the powers of 2 being multiplied by ones.

The card that has number 1 on top left represents the "digit" f. The card that has number 2 on top left represents the "digit" e. The card that has number 4 on top left represents the "digit" d. And so on. If a person says yes, you put 1 at the position, if the person answers no, put a zero. For example, if a person thinks of the number 37, then we consider its binary representation: $37 = 32 + 4 + 1 = (100101)_2$, then number 37 will only appear on the first, third and last card (the sequence of cards is opposite to the digit position).

Magic Trick 02: Trick using Base 3

Material: You will need 27 cards of a regular deck of cards.

Preparation: You will need to memorize the table on Appendix 2.

Performance: Ask someone to look at the 27 cards and mentally choose one. Then the person shuffles the cards. While shuffling, ask the person to choose a number between 1 and 27, do not make a big deal out of this number at this point. Take all cards face down on your hands and start to deal the cards one-at-a-time face-up creating three piles of cards. Deal cards from left to right, one card per pile. After all cards are dealt, ask the person to point which pile the chosen card is. Pick the three piles (that's when the magic happens). Repeat the process of distributing the cards into three piles and asking which pile the card is. Pick again the piles (more magic happening). And repeat the process once more: dealing, asking the pile, picking up piles.

Give all the cards to the person, and ask again which number he/she had chosen before. Then, ask the person to count that amount of cards from the top. The chosen card will be exactly the card appearing on the position of the chosen number. For a nice effect, right before the person flips the very last card, ask the person to say out loud the card.

Trick: It is a more elaborated trick. It consists on how you stack up the piles after the person points to the pile. The way you do will depend on the number that the person first chose. See appendix 2. Basically, there are three possibilities that you may put the pile containing the chosen card:

- T: You could put that pile on <u>T</u>op of the other two piles;
- M: You could put that pile on the <u>M</u>iddle of the other two piles;
- B: You could put that pile on the <u>B</u>ottom of the other two piles.

On each step, you will put the pile with the card on T, M, or B, according to the table on appendix 2.

Explanation: Notice that if the card is in position n, that means that there would be $n-1$ cards on top of it. We will take into consideration how many cards it is needed to be removed before the chosen card is revealed. That means, if the person says number 16, we will work with the number 15.

There are 27 cards: $27 = 3^3$. There are 27 possible numbers that can be written using three-digits on a base 3 system: from $0 = (000)_3$ to $26 = (222)_3$. Here is a table with all conversions:

0	000	9	100	18	200
1	001	10	101	19	201
2	002	11	102	20	202
3	010	12	110	21	210
4	011	13	111	22	211
5	012	14	112	23	212
6	020	15	120	24	220
7	021	16	121	25	221
8	022	17	122	26	222

The way you stack the piles for the first time, you are either putting the cards:

- Top: meaning the chosen card is one of the first nine cards (positions 1 to 9);
- Middle: chosen card's position is between 10 and 18;
- Bottom: chosen card is in a position between 19 and 27.

That is, on your first move, you are choosing the left-most digit on the three-digit base-3 number that represents how many cards are to be removed. Top would be zero, middle is one, and bottom is two.

Then, you deal again the cards. Depending on how you stack the piles on the previous step, the chosen card will be either on the top, middle, or bottom third of a pile with nine cards. And the way you stack the piles on the second time determines the position of the card within the 27 cards. Depending on whether your first choice was top, middle or bottom, you may have only a group of three possible cards:

- Top: position 1 through 3, or 10 through 12, or 19 through 21.
- Middle: 4 through 6, or 13 through 15, or 22 through 24.
- Bottom: position 7 through 9, or 16 through 18, or 25 through 27.

That is, on your second move, you are choosing the middle digit on that three-digit base-3 number representing how many cards are to be removed. Again: Top would be zero, middle is one, and bottom is two.

Finally, on your third time, you are basically pin-pointing for each group of three possible cards (after the second step) which card you are selecting. Then, you stack according to its order:

- Top: means that the chosen card is the first one from the three-card group, that is, according to previous choices, position is: 1, 4, 7, 10, 13, 16, 19, 22, or 25.
- Middle: means that the chosen card is the second one from the three-card group, that is, according to previous choices, position is: 2, 5, 8, 11, 14, 17, 20, 23, or 26.
- Bottom: means that the chosen card is the third one from the three-card group, that is, according to previous choices, position is: 3, 6, 9, 12, 15, 18, 21, 24, or 27.

That is, on your third move, you are choosing the right-most digit on that three-digit base-3 number representing how many cards are to be removed. Again: Top would be zero, middle is one, and bottom is two.

Example: Suppose the person says number 16. From the table, 16 is "T B M" (top, bottom, middle). Ask the person to choose a card, deal the cards into three piles, one card at a time on each pile. When the person points which pile the chosen card is, stack the three piles, keeping the pile with the card on the top.

Then, deal again, and when the person points out the pile on this second step, stack that pile on the middle.

Finally, the chosen pile on the third step will be on the bottom.

Done! The chosen card should be on position 16.

Hint: Practice few times so your moves seem natural. Avoid showing that you are thinking about which pile to pick first.

Magic Trick 03: Another Trick using Base 3

Material: Blank paper and a pen or pencil.

Preparation: You will need to learn and rehearse with a friend the contents of the table on Appendix 3. Place a friend far from you. Your friend should not be looking at you.

Performance: You will pretend to communicate with a friend telepathically. Ask someone to choose a number with 6 to 8 digits and write on a piece of paper. You look at the number, but your friend doesn't. You concentrate on the number, pretending you are setting a Wi-Fi connection with your friend.

1) Your friend asks whether you are ready. You confirm that you are.
2) Your friend, without looking at you, guesses the first number. You confirm it is correct.
3) Your friend guesses one-by-one, and you confirm one-by-one.

Trick: The secret to this trick is that you are telling the numbers one-by-one when you give confirmation to the previous question. You will answer the first question according to which the first digit is. You may say:

- Ok, ok: if the number is 0;
- Ok, yeah: if the number is 1;
- Ok, yes: if the number is 2;
- Yeah, ok: if the number is 3;
- Yeah, yeah: if the number is 4;
- Yeah, yes: if the number is 5;
- Yes, ok: if the number is 6;
- Yes, yeah: if the number is 7;
- Yes, yes: if the number is 8;
- Or any other thing, like "it is correct" or "great" or "got it", if the number is 9.

The table on appendix 3 summarizes this. Notice that it is organized in alphabetical order. If the first word is "Ok", then the number is between 0 and 2. If the first word is "Yeah", then the number is between 3 and 5. And if the first word is "Yes", number will be from 6 to 8.

Explanation: We consider the base 3 representation of all numbers from 0 to 8:

$0 = (00)_3$ $3 = (10)_3$ $6 = (20)_3$

$1 = (01)_3$ $4 = (11)_3$ $7 = (21)_3$

$2 = (02)_3$ $5 = (12)_3$ $8 = (22)_3$

And we identify each word with a digit on ternary (base-3) system: Ok $= 0$, Yeah $= 1$, Yes $= 2$.

So, "Ok, ok" is the same as $(00)_3$, "Ok, yeah" is $(01)_3$, and so on.

Hint: The trick is simple, but it requires rehearsal. Practice with a friend several times before showing this trick. The person who is "sending the signal" needs to have the confirmation already in mind before the person says the guessing number, so it looks natural. Do not repeat this trick to the same audience, since with repetition, they will start to notice that the "ok, yeah, yes" pattern changes with answer.

To sound more natural, every time that a word is supposed to be repeated just say it once. In this case, if the confirmation is "Ok, ok" simply say "Ok" then the number should be 0, "Yeah" refers to 4, and "Yes" to 8. For memorization, one-words refer to multiples of 4.

<u>Example:</u> Suppose a person (audience) writes the number 618903. You will pretend to send this number to a friend.

You mentally go over each digit and prepare the right expressions. Take your time, since this step is important, especially if you are performing the trick in the first times. Your friend starts by asking something like:

    Friend: Can we start?

Since the first digit is 6, you answer:

    You: Yes, ok.

In the meantime, you already prepare the answer "ok yeah" (since it is number 1) for the next number.

Friend takes a bit to answer, so you have time to have the next answer prepared and says:

    Friend: I think the first number is 6, is it?

    You: Ok, yeah.

And this should be repeated. Friends delays the answer, you prepare the next confirmation.

    Friend: Number 1. Did I get it right?

    You: Yes.

    Friend: Is it 8?

    You: Affirmative.

    Friend: Is 9 the next digit?

    You: Ok.

    Friend: How about number 0?

    You: Yeah, ok.

    Friend: 3.

Magic Trick 04: A Trick using Base 5

Material: 25 cards from a regular deck.

Preparation: You will need to somehow memorize the first five cards on top of the deck without being caught. As a hint, you may use a nice sequence on top, for instance: Ace, 3, 5, 7, and 9, of a specific suit. It may require some shuffling tricks, for instance, you have put the cards on top, but then you shuffle the cards making sure that the top five cards are not changed. That movement may require some practice.

Performance: Divide the cards into five groups of five cards by putting the cards "by layers". That is,

1) First put five face-down cards to be the opening on each pile.
2) Then, put another face-down card on top of each card until all piles have two cards. The sequence in which you put the layers may vary, i.e., you don't need to repeat the same pattern of piles when putting the next layer, but you have to finish a 5-card-layer before starting the next one.
3) Distribute all 25 cards into the 5 piles, by repeating the "by layer" process.
4) Once all cards are distributed, ask someone to choose a pile, look at the five cards, and mentally choose one of them.
5) Ask this person to shuffle each pile separately. That is, the person may change the order of cards within each pile, on all five piles. But he/she cannot transfer cards between piles.
6) Ask the volunteer to make a single pile of cards by putting one pile on top of the others, in any order.
7) Then, ask the person to "cut" the pile by dividing it into two and putting the bottom pile on top.
8) Now, you divide the cards by taking five consecutive from the deck cards and putting them face-up. Repeat until you have five groups of five face-up cards.
9) Ask which group the chosen card is.
10) Once you hear it, you immediately say the chosen card.

Trick: You have to memorize the first five cards on top of the deck. When you distribute the cards, notice that each of the memorized cards will be part of a different pile. Now you have a "key" card to focus on, which is the card from that chosen pile that you already know.

Once you distribute again the card in layers, each card of that pile from previous step will be on a different new pile. And notice that they will all have the same relative position, in other words, the cards from that chosen pile will either all be the first of each new pile, or the second, or third, or fourth, or fifth.

But you have the "key" card, so you know exactly which relative position the cards appears within each of the new piles.

That's when you immediately determine the card.

Explanation: The secret to this trick is that the cards are distributed in a "base 5" configuration. By memorizing the top card, you know a digit on a two-digit base 5 number. The last digit is then determined when the volunteer chooses the pile at the end of the trick. Remember that with two-digits, there may be 25 different combinations:

$(00)_5 = 0$ $\quad\quad$ $(01)_5 = 1$ $\quad\quad$ $(02)_5 = 2$ $\quad\quad$ $(03)_5 = 3$ $\quad\quad$ $(04)_5 = 4$

$(10)_5 = 5$ $\quad\quad$ $(11)_5 = 6$ $\quad\quad$ $(12)_5 = 7$ $\quad\quad$ $(13)_5 = 8$ $\quad\quad$ $(14)_5 = 9$

$(20)_5 = 10$ $(21)_5 = 11$ $(22)_5 = 12$ $(23)_5 = 13$ $(24)_5 = 14$

$(30)_5 = 15$ $(31)_5 = 16$ $(32)_5 = 17$ $(33)_5 = 18$ $(34)_5 = 19$

$(40)_5 = 20$ $(41)_5 = 21$ $(42)_5 = 22$ $(43)_5 = 23$ $(44)_5 = 24$

For simplicity, let's assume that the cards are distributed on piles next to each other side-by-side, so we can enumerate them: Pile 0, Pile 1, Pile 2, Pile 3, and Pile 4. And within each pile, we also enumerate the cards: Card 0, Card 1, Card 2, Card 3, and Card 4. Let's use 0 to be the first card. Hence, Pile 0 / Card 0 means the first card.

When the volunteer picks the card by first looking into a pile, he/she is choosing the right digit on the two-digit base 5 number. Finally, by indicating the pile where the card is, he is identifying the left-digit, so the number is determined, and you can identify the card.

Hint: The trick is simple, but the preparation requires rehearsal on how to make sure that it looks like you are shuffling the cards, while the top five cards are kept in the same order.

Read again these instructions with 25 cards to see step-by-step how the trick is performed.

Appendices

Appendix 1: Binary cards

Appendix 2: Trick using base 3

Appendix 3: Another trick using base 3

Appendix 1 – Binary cards

1	3	5	7	9	11	13	15
17	19	21	23	25	27	29	31
33	35	37	39	41	43	45	47
49	51	53	55	57	59	61	63

2	3	6	7	10	11	14	15
18	19	22	23	26	27	30	31
34	35	38	39	42	43	46	47
50	51	54	55	58	59	62	63

4	5	6	7	12	13	14	15
20	21	22	23	28	29	30	31
36	37	38	39	44	45	46	47
52	53	54	55	60	61	62	63

Appendix 1 – Binary cards

8	9	10	11	12	13	14	15
24	25	26	27	28	29	30	31
40	41	42	43	44	45	46	47
56	57	58	59	60	61	62	63

16	17	18	19	20	21	22	23
24	25	26	27	28	29	30	31
48	49	50	51	52	53	54	55
56	57	58	59	60	61	62	63

32	33	34	35	36	37	38	39
40	41	42	43	44	45	46	47
48	49	50	51	52	53	54	55
56	57	58	59	60	61	62	63

Appendix 2 – Trick using base 3

1	-	T	T	T
2	-	M	T	T
3	-	B	T	T

10	-	T	T	M
11	-	M	T	M
12	-	B	T	M

19	-	T	T	B
20	-	M	T	B
21	-	B	T	B

4	-	T	M	T
5	-	M	M	T
6	-	B	M	T

13	-	T	M	M
14	-	M	M	M
15	-	B	M	M

22	-	T	M	B
23	-	M	M	B
24	-	B	M	B

7	-	T	B	T
8	-	M	B	T
9	-	B	B	T

16	-	T	B	M
17	-	M	B	M
18	-	B	B	M

25	-	T	B	B
26	-	M	B	B
27	-	B	B	B

Appendix 3 – Another Trick using base 3

Second: \ First:	Ok	Yeah	Yes
Ok	0	3	6
Yeah	1	4	7
Yes	2	5	8
Something else: 9			

About the author

Ricardo Teixeira, Ph.D.
University of Houston – Victoria
Director of Math Program
Director of Core Curriculum
Assistant Professor of Math

Ricardo Teixeira began teaching at the University of Houston – Victoria (UHV) in the fall of 2010, when the university received the first class of freshman and sophomore students. He helped implemented the lower-division math courses and first-year seminar. Since then, Dr. Teixeira has been teaching lower-division, upper-division and graduate courses at UHV.

Research Interests

Dr. Teixeira has interests in both pure and applied Mathematics, as well as in Mathematics Education. In pure Mathematics, his main work is in the Functional Analysis field. He received his PhD from The University of Texas at Austin, advised by Dr. Edward Odell, in the spring of 2010. Dr. Teixeira solved the problem of whether the set of S_α-Singular operators form an ideal, this work was published in the Proceedings of the American Mathematical Society. Currently, he studies structure of infinite-dimensional Banach Spaces.

In applied Mathematics, Dr. Teixeira has worked in developing tools to predict the results of future events. His tool can be successfully applied to predicting the outcome of sports events. Under this research line, Dr. Teixeira has already advised two graduate students in projects. As future projects, he is trying to expand the use of predictions to marketing and other applications.

In Mathematics Education, Dr. Teixeira has contributed to researches regarding how college students may mature quantitative reasoning abilities. He has also worked with Dana Center in developing textbooks for an innovative approach to developmental/freshman level mathematics. The project is called New Mathways Project, already implemented in most Community Colleges in Texas and in several other institutions across the US, uses the most modern ideas (learning episodes, inquiry-based techniques, real-data problems, etc.) to present mathematics in an interesting and engaging way.

Supporting Mathematics

Dr. Teixeira also supports education through different lines. He is the faculty advisor of an honor society (UHV's Gamma Beta Phi), the founder of a science club for elementary school students, the director of the Mathematics Program at UHV, the director of the Core Curriculum at UHV, the organizer of the annual Math and Robotics Awareness Day (event that has brought several hundreds of high school students to campus for activities and competitions), a member in the Living and Learning Communities, a member of the STEAM panel at Children's Discovery Museum, a member of the Site-Based Decision Team for a local elementary school, and member of several other committees in Victoria.

www.ingramcontent.com/pod-product-compliance
Lightning Source LLC
Chambersburg PA
CBHW081209180526
45170CB00006B/2279